Preface:

This book is a condensed explanation of frequency mixers used in the electronic communication industry. Chapter 1 starts by defining the mixer and introduces concepts such as the harmonic intermodulation tables, input and output intercept points, and other definitions. Chapter 2 introduces more advanced concepts such as BPSK and QPSK decoding and the Weaver method of single sideband generation and demodulation. Chapter 3 is a full derivation of equations that define the gilbert cell, considering that most integrated circuits that implement mixers do so by using a gilbert cell. Chapter 4 is an overview of different types of mixers, starting with the Pentagrid mixer, includes different types of mixers used in analog and digital phase locked loop circuits, and ends in a little know technique using analog-to-digital conversion as a mixer for down converting. Chapter 5 shows how to use mixers for analog computation functions such as multiplication, division, squaring, and square root, including powers and roots that are non-integral. Chapter 6 gives applications with working schematic slices for each example. Chapter 7 provides complete schematic examples of different types of mixers used in radio receivers.

Contents

CHAPTER 1

Introduction to Mixers

Mixers are used for frequency conversion and are critical components in modern radio frequency (RF) systems. A mixer converts RF power at one frequency into power at another frequency to make signal processing easier and also inexpensive. A fundamental reason for frequency conversion is to allow amplification of the received signal at a frequency other than the RF, or the audio, frequency. A receiver may require as much as 140 decibels (dB) of gain. It might not be possible to put more than 40 dB of gain into the RF section without risking instability and potential oscillations. Likewise the gain of the audio section might be limited to 60 dB because of parasitic feedback paths, and micro-phonics. The additional gain needed for a sensitive receiver is normally achieved in an intermediate frequency (IF) section of the receiver

The ideal mixer, represented by Figure 1, is a device which multiplies two input signals. If the inputs are sinusoids, the ideal mixer output is the sum and difference frequencies given by

$$V_0 = [A_1\cos(\omega_1 t)][A_2\cos(\omega_2 t)] = \frac{A_1 A_2}{2}[\cos(\omega_1 - \omega_2)t + \cos(\omega_1 + \omega_2)t]$$

Typically, either the sum, or the difference, frequency is removed with a filter.

Figure 2 shows the front end for an amplitude modulation (AM) super heterodyne radio with a tuned RF section that allows only the preselected RF frequency to the input of the mixer. Figure 3 improves the front end image rejection by adding an additional tuned RF circuit. Assuming the desired station is transmitting at 1490 KHz, with a high side local oscillator (LO) of 1945 KHz, and an IF of 455 KHz, then a station transmitting at 2400 KHz would also convert to an IF of 455 KHz, interfering with the desired station, if the 2400 KHz signal were not filtered out prior to the mixer input. In a typical AM radio, the front end is tuned simultaneously with the LO at a constant difference of 455 KHz over the entire AM radio band of 550 - 1600 KHz.

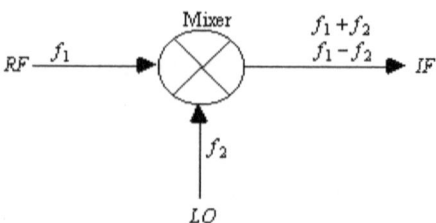

Figure 1: Basic Mixer Operation

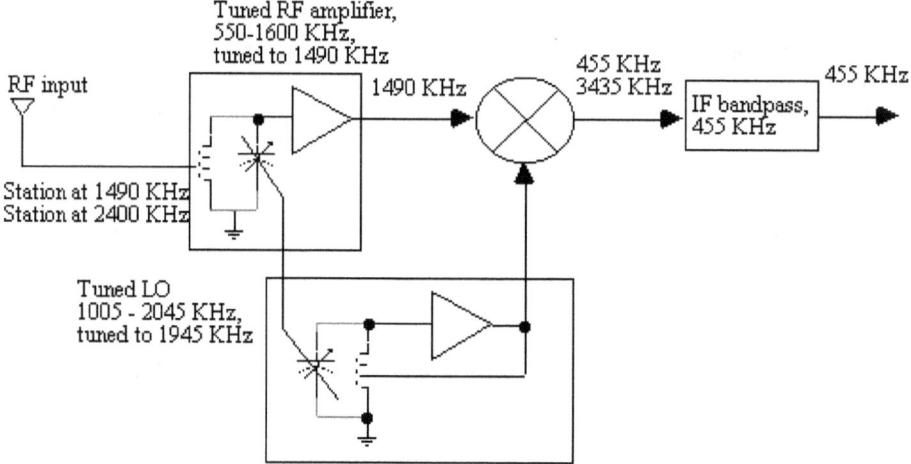

Figure 2: Super heterodyne radio front end

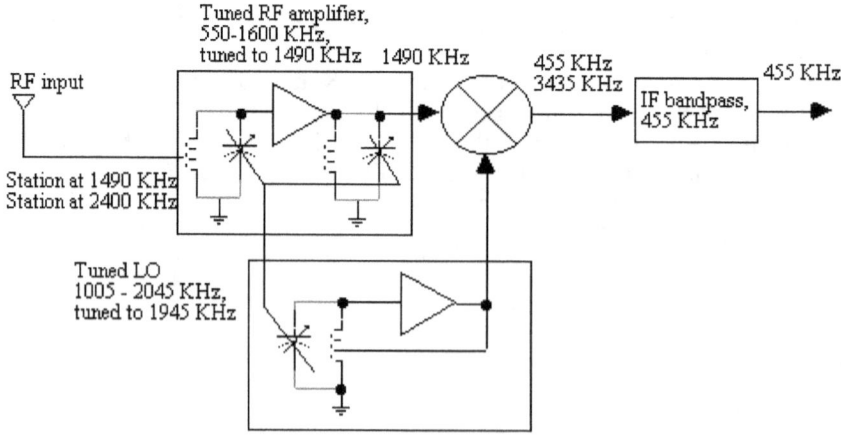

Figure 3: Super heterodyne AM radio front end with improved front end filtering

This requires a specially constructed dual ganged-variable capacitor to tune both sections simultaneously and retain a constant 455 KHz difference. Typically the LO section of the ganged-variable capacitor will have fewer plates than the RF amplifier section. This type of ganged-variable capacitor is called the *cut-plate* capacitor because the LO section plates are cut to permit tracking of the LO with the RF. The undesired frequency of 2400 KHz is called the image frequency and must be eliminated prior to conversion. The 455 KHz IF filter eliminates the sum frequency of 3435 KHz which contains redundant information.

FM radios, which tune over 88 - 108 MHz, usually do not use a 455 KHz IF frequency since the image frequency would be only 910 KHz from the desired FM station. It would be difficult to design a tuned RF amplifier in the 88 - 108 MHz range that rejected a station only 910 KHz away from the desired signal since 910 KHz is only about 1% different in frequency than the desired FM station. An IF of 10.7 MHz is normally used in FM radios to allow adequate image rejection to be achieved by the tuned RF amplifier in the 88 - 108 MHz band with reduced selectivity because of the higher IF bandwidth associated with a 10.7 MHz IF filter. The double conversion receiver of Figure 4 is appropriate for VHF narrowband AM or FM operation and uses a 10.7 MHz 1st IF for good image rejection, and a 455 KHz 2nd IF for good selectivity.

The double conversion receiver in Figure 5 does not require a tunable RF stage to track the tunable LO because the first IF of 830 MHz causes the image frequencies to be outside the band pass of the 0-30 MHz input filter. This technique of commonly used in communication grade receivers, CATV tuners, and spectrum analyzers.

The direct conversion receiver of Figure 6 suffers from several disadvantages. It does not have an intermediate frequency (IF) stage. The purpose of an IF stage is to allow additional amplification at a non-harmonically related frequency that will not feed back into the RF input and cause oscillation. The gain of the direction conversion receiver is therefore limited to the gain of any RF amplifiers preceding the mixer, and any audio amplifiers following the mixer. AM and SSB can be demodulated, but not FM. There is usually significant LO feed through at the desired frequency which can cause undesired beats in the audio output for AM signals.

In reality, mixers produce more than just the sum and difference frequencies. The intermodulation products are given by IF = N*R ±M*LO and their levels, relative to the desired output of RF ± LO, for a common mixer, are shown in table 1.

3

In table 1, the spurious outputs are relative to the desired output at RF ± LO. The first harmonic of the RF input signal feeds through to the output and is only 23 dB down from the desired output frequency. The first harmonic of the LO feeds through at only 1 dB down. The second harmonic of the RF input, mixed with the first harmonic of the LO (2RF ± LO), feeds through at an amplitude of 59 dB down from the desired output, and so on.

The data given in a harmonic intermodulation table depends on the relative levels of the input signals, the frequencies, and the terminating impedances. From table 1, notice that the SRA-200 mixer is a poor choice for the direct conversion receiver since the LO leakage is high. Also notice that the front end of the receiver in Figure 5 is not tuned, and will also have some response to other RF frequencies as shown in table 1 where 2 RF2 + LO, 3 RF3 + LO, etc., are equal in frequency to the desired signal at frequency RF + LO.

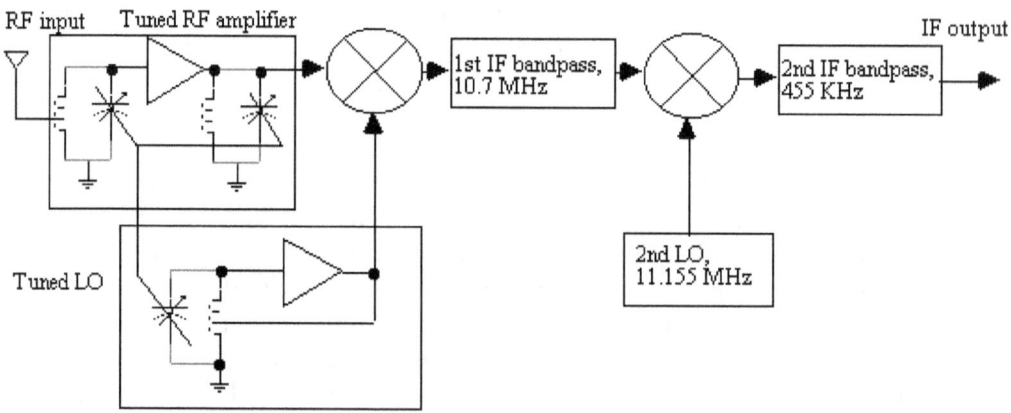

Figure 4: Double conversion super heterodyne with good image rejection and selectivity

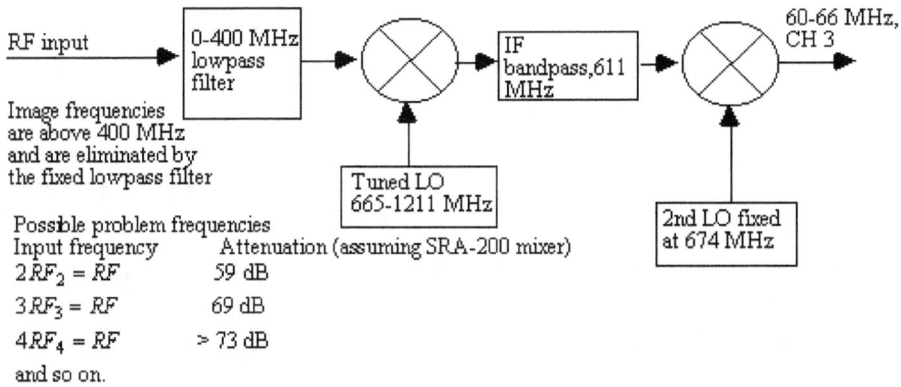

RF input

0-400 MHz lowpass filter

IF bandpass,611 MHz

60-66 MHz, CH 3

Tuned LO 665-1211 MHz

2nd LO fixed at 674 MHz

Image frequencies are above 400 MHz and are eliminated by the fixed lowpass filter

Possible problem frequencies

Input frequency	Attenuation (assuming SRA-200 mixer)
$2RF_2 = RF$	59 dB
$3RF_3 = RF$	69 dB
$4RF_4 = RF$	> 73 dB

and so on.

Figure 5: Double conversion receiver in a Scientific-Atlanta series 6700 cable converter box that eliminates the need for simultaneously tuned RF and LO

Figure 7 shows a diode ring double-balanced mixer which is usually designed with Schottky barrier diodes. GaAs diodes are sometimes used for operation in the millimeter-wave frequency range. Mixers can also use bipolar transistors, J-FETs, and GaAs FETs, all of which require a fourth port for a dc voltage.

For down converters, the RF input signal is fed to the RF port and the output is taken from the IF port. For upconverting applications using the diode ring double-balanced mixer, the low frequency signal can be fed into the IF port and the output can be taken from the RF port.

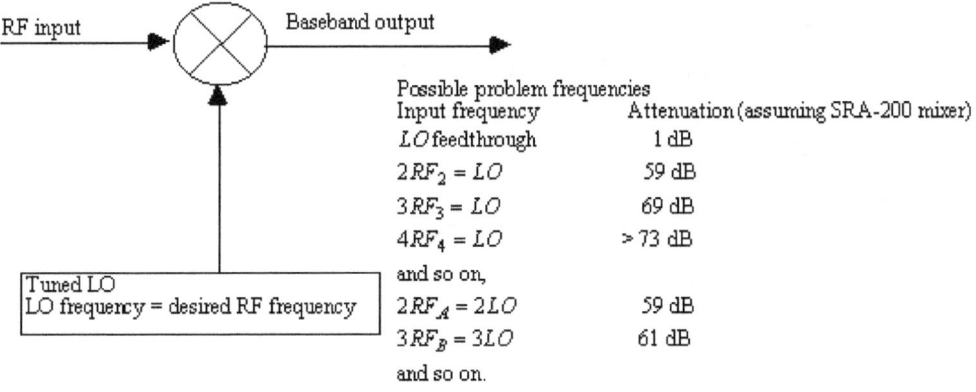

RF input

Baseband output

Tuned LO
LO frequency = desired RF frequency

Possible problem frequencies

Input frequency	Attenuation (assuming SRA-200 mixer)
LO feedthrough	1 dB
$2RF_2 = LO$	59 dB
$3RF_3 = LO$	69 dB
$4RF_4 = LO$	> 73 dB
and so on,	
$2RF_A = 2LO$	59 dB
$3RF_B = 3LO$	61 dB

and so on.

Figure 6: Direct conversion receiver

Table 1

Mixer Harmonic Intermodulation for Mini-Circuits SRA-220, 0.05 - 2000 MHz mixer

```
                Harmonic LO Order
                0    1    2    3    4    5    6    7    8    9   10
R   0    –    –    1   16   16   31   52   33   28   68   49   63
F   1    –   23    –   29   23   30   42   43   38   56   35   48
    2  >95   61   59   59   61   66   71   70   72   64  >67   67
O   3  >94  >72   69  >74   63  >73   72  >71  >74  >74  >74   67
r   4  >96   75  >73  >73  >74  >74  >74  >74  >74  >74  >73  >75
d   5  >97  >74  >74  >73  >73  >72   75  >73  >72  >74  >74  >73
e   6  >90  >68  >74  >74   73  >73  >73  >75  >73  >73  >74  >74
r   7  >89  >67  >67  >74  >75  >75   74  >74  >74  >75  >73  >73
    8  >90  >67  >69  >67  >73  >74  >74  >73  >73  >73  >74  >74
H   9  >89  >68  >68  >68  >67  >74  >74  >73  >74  >72  >74  >73
a  10  >91  >68   68  >68  >68  >68  >74  >74  >74  >71  >74  >74
r       0    1    2    3    4    5    6    7    8    9   10
m
o            Harmonic LO Order
n
i
c
```

Test conditions: RF = 999.1 MHz, Drive level = -13.99 dBm
LO = 969.01 MHz, Drive level = 10 dBm
IF = 30.095 MHz, Measured IF level = -21.81 dBm

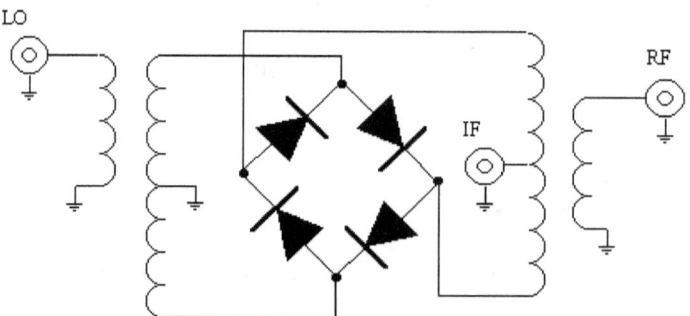

Figure 7: Diode ring double-balanced mixer

Definitions

Conversion Loss is the ratio of the output signal level to the input signal level expressed in dB. In a single sideband system, only one sideband is used; therefore 3 dB of loss is theoretical. The additional loss is diode and transformer loss. These losses can be minimized by driving the diodes with sufficient current and operating in the best portions of the frequency band and are generally between 5 - 9 dB for passive mixers. Conversion loss is specified in a 50 Ω system with an LO drive level of +7 dBm. High level mixers are specified with more LO drive power.

Isolation is the amount of "leakage" or "feedthrough" between the mixer ports. From table 1 for the SRA-220 mixer, the isolation between the LO port and the IF (output) port is only 1 dB, for the test conditions given. At low frequencies, where diode parameters can be matched to a much greater degree and circuit parasitics are negligible, isolation greater than 60 dB is possible.

Noise Figure is the signal-to-noise ratio at the input divided by the signal-to-noise ratio at the output expressed in dB. It does not include the noise figure of an IF amplifier or 1/f flicker noise. Appreciable noise contribution from 1/f noise is not noticeable above 10 KHz. Use of specially selected Schottky-Barrier diodes ensures extremely low 1/f noise for phase detection applications. With the recommended drive level, the noise figure and conversion loss are essentially identical.

The Friis noise figure equation for cascaded devices is given as

$$F = F_1 + \frac{F_2 - 1}{G_1} + \frac{F_3 - 1}{G_1 G_2} + \frac{F_4 - 1}{G_1 G_2 G_3}$$

where F is the total noise factor. All terms are numeric ratios and are not in dB. The overall noise figure for a cascade, expressed in dB is

$$F_{dB} = 10 * \log(F)$$

If a manufacturer of a RF preamplifier specifies that the RF preamp has a 2 dB noise figure, this means that the actual noise power at the output is 1.58 (F = 1.58) times that which would occur due to amplification of the thermal noise from the input. If a device is passive and lossy then the noise figure of that device is generally taken to be the same as its loss. If other words, if a transmission line has a 3 dB loss, it has a noise figure of 3 dB.

Conversion Compression is the RF input level above which the RF versus IF output curve deviates from linearity. Above this level, additional increases in the RF input level do not result in equal increases in the output level. Conversion compression is not specified for all low level (+7 dBm LO) mixers. However, low level mixers normally have the same compression level, typically 0.3 dB deviation from linearity with an RF input signal of +2 dBm and a +7 dBm LO drive level. This RF level can be raised to +4 dBm if the drive level is increased to +13 dBm. Conversion compression for high level mixers is specified since it sometimes provides an indication of the mixer's two tone performance and it is likely to be important in high level operation. The input power level at which the conversion loss increases by 1 dB is called the 1 dB compression point.

Dynamic Range is the amplitude range over which a mixer can operate without degradation of performance. It is bounded by the conversion compression point for high input signals, and by the noise figure of the mixer for low level input signals. Since the thermal noise of each passive mixer is about the same, the conversion compression point normally determines the passive mixer's dynamic range. The 1 dB compression point is generally taken to be the top of the dynamic range of a mixer because the input RF power that is not converted into desired IF output power is instead converted into heat and higher order intermodulation products. The intermodulation products that begin to appear when RF power is increased beyond the 1 dB compression point can begin to obscure the desired IF output. Generally the 1 dB compression point is 5 to 10 dB lower than the LO power, so a high level mixer has a higher 1 dB compression point than a low level mixer, and therefore a wider dynamic range.

Intercept Point, measured in dBm, is a figure of merit for intermodulation product suppression. A high intercept point is desirable. Two types are commonly specified: input and output intercept point (IIP and OIP, respectively). Input intercept point is the level of input RF power at which the output power levels of the undesired intermodulation products and IF products would be equal; that is, intercept each other if the mixer did not compress. As input RF power increases, the mixer compresses before the power level of the intermodulation products can increase to equal the IF output power. So, input and output intercept points are theoretical and are calculated by extrapolating the output power of the intermodulation and IF products past the 1-dB compression point until they equal each other. A high intercept point is desirable because it means the mixer can handle more input RF power before causing undesired products to rival the desired

IF output product, and essentially means the mixer has a greater dynamic range. Dynamic range, 1-dB compression point, and intercept point are all interrelated, but it has been shown that, in general, no dB-for -dB rule of thumb exists to easily correlate 1-dB compression point with intercept point.

The concept of intercept point can be applied to any intermodulation product; however, it normally refers to two-tone, third-order intermodulation products. If two input RF signals are incident at the mixer RF port, they cause the mixer to generate the following multi-tone intermodulation products, where m1, m2, n = 0, 1, 2, 3,..., m and n are integers and can assume any value. Two-tone, third-order intermodulation products have the following frequencies:

$$(\pm RF_1 \pm 2RF_2) \pm LO$$

and

$$(\pm 2RF_1 \pm RF_2) \pm LO$$

They are called third-order products because the coefficients of RF1 and RF2 sum to equal 3. Notice that the order of intermodulation products refers only to coefficients of the RF inputs and does not include that of the LO. The order of the intermodulation product is important because a 1-dB change in the power level of each input RF signal causes the power level of each intermodulation product to change by an amount of dB equal to its order. A 1-dB change in power of each of the two input RF signals causes the power level of each tow-tone third-order product to change by 3 dB.

Input intercept point is normally associated with tow-tone, third-order intermodulation products because the third-order product is closest in frequency to the desired IF output product of any tow-tone intermodulation product. The even-order, two-tone intermodulation products that exit form double and single-balanced mixers are suppressed far more than the odd-order products, due to mixer balance. Odd order intermodulation products containing even order LO harmonics are suppressed in double, but not in single balanced mixers. Third-order two-tone products follow the (m1 + m2) dB of output power to 1-dB of input-power-rule much more closely than the other higher-order, two-tone intermodulation products. Two-tone intermodulation products with orders greater than 7 are rarely a problem unless RF input power comes within a few dB of LO input power.

Intercept point in normally presented as shown in Figure 8. Input power is plotted along the horizontal axis, and output power is plotted along the vertical axis. Two lines are plotted: one relating IF output power to RF input power, and another relating intermodulation output power to RF input power.

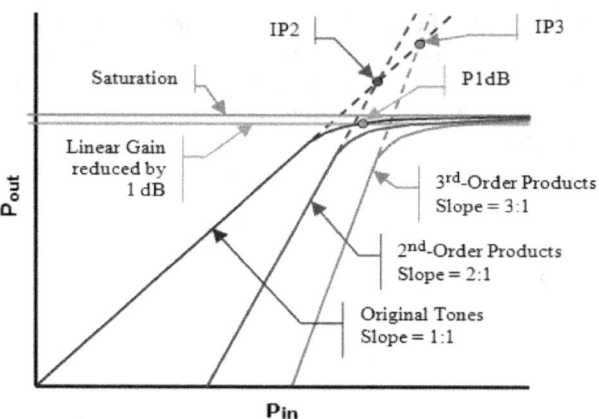

Figure 8: Intercept point definition

The point at which these lines intersect gives the input and output intercept points for the mixer at a particular set of input frequencies for a given LO power level and temperature.

A simple formula exists for calculating input intercept point, given the level of intermodulation suppression, the order of the intermodulation, and the input RF power levels giving rise to this level of suppression.

$$IIP = \left[\frac{\text{Intermodulation suppresion(dBc)}}{(\text{order} - 1)} \right] \pm [\text{input RF power(dBm)}]$$

For example, if each input tone has -10 dBm of power, and the third-order, two-tone intermodulation suppression is 46 dBc, then the IIP is

$$IIP = 46 \frac{46}{3-1} + (-10dBm) = +13dBm$$

Also, output and input intercept are related by the mixer conversion loss, or gain (for active mixers).

$$OIP(dBm) = IIP(dBm) \begin{bmatrix} - \text{mixer conversion loss(dB), or} \\ + \text{mixer conversion gain(dB)} \end{bmatrix}$$

Voltage Standing Wave Ratio (VSWR) is the measure of mismatch offered to the system by the mixer, and is usually specified over a given bandwidth as a function of LO power and temperature. It is calculated as

$$VSWR = \frac{1 + |\rho|}{1 - [\rho]}$$

where
$$\rho = \frac{Z_L - Z_o}{Z_L + Z_o}$$

ρ is the reflection coefficient,

Z_L is the input impedance of the mixer, and

Zo is the characteristic impedance of the system.

Since VSWR does not include the phase of the reflection coefficient, the system designer does not know if the input impedance is above or below the normal 50 Ω characteristic impedance. For example, if the LO port VSWR is 2:1, measured in a 50 Ω system, the designer does not know if the LO port input impedance is 25 Ω or 100 Ω since both these impedances give a VSWR of 2:1. Actually, the input impedance of a broadband mixer swept over a frequency range of an octave, or more, usually rotates through the low and high impedances, roughly producing a circle centered at 50 Ω as viewed on a Smith chart. Therefore, a given mixer having a LO VSWR of 2:1 over an octave bandwidth will have an input impedance varying from 25 Ω to 100 Ω, passing through an infinite number of complex impedance combinations as the LO frequency changes. The VSWR of the RF, LO, and IF ports are direct functions of the LO power, which establishes the operating point of the diodes in a diode ring mixer. Change in the LO power alters the diode operating point, resulting in a different impedance of all mixer ports, causing a corresponding change in VSWR. RF input power, which is usually much lower than LO input power, does not appreciably change the diode bias point and consequently, has little affect on VSWR. When the diode impedance changes, the input impedances of all three ports change. Hence, varying the LO power level will affect the VSWR of all three ports.

CHAPTER 2

The Linear Diode Detector for AM Demodulation

An ideal diode conducts only during alternate half cycles of the input signal, and during the conducting half cycles the output current is proportional to the input voltage. An AM signal, applied to a diode detector as shown in Figure 9, reproduces the modulating (audio) signal by mixing the AM sidebands with the AM carrier. It may be seen from Figure 1 that the peak amplitude of each current pulse in the output is proportional to the peak amplitude of the input voltage during that particular conducting half cycle. Thus the peak, and therefore the average, values of the output current pulses follow the amplitude of the input voltage precisely during conducting half cycles and have the same waveform as the modulation envelope. Whether the output voltage would approach the peak or average value of the input voltage depends on the type of filter used in the output. If the filter is a bypass capacitor, and the internal resistance of the rectifier is small in comparison with the load resistance, then output voltage will tend to follow the peaks of the input voltage. Thus, the capacitor, CF, charges essentially to the peak input voltage during the conducting half cycles, but there is not time for appreciable discharge through the high resistance load during the non-conducting half cycles. If the bypass capacitor is too large, the time constant of the discharge will be so large that the detector output will not be able to follow the modulation envelope when the modulation envelope decreases in amplitude rapidly.

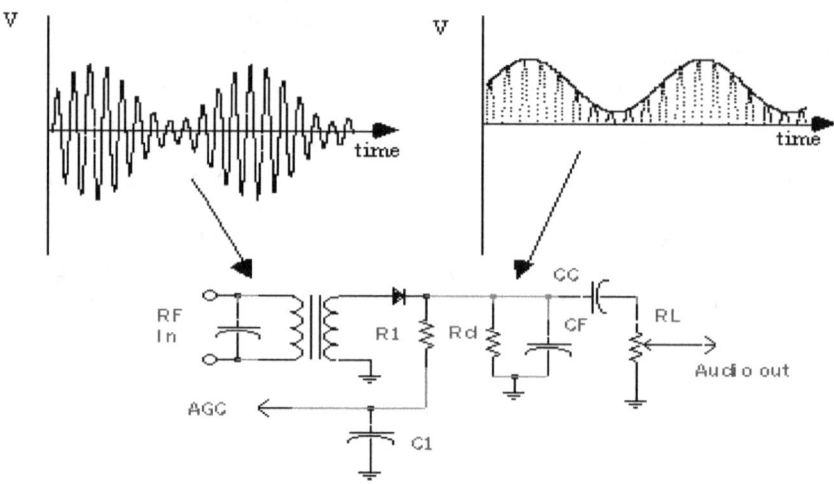

Figure 9: AM diode detector

In addition to detection, a circuit in Figure 9 has been added which is called the automatic gain control (AGC). The AGC voltage is the average value of the detector output voltage since R1 and C1 act as a filter to remove the modulating signal as well as the RF from the AGC system. This AGC voltage is therefore proportional to the amplitude of the carrier in a continuous wave system and may be used to automatically control the gain of one or more RF amplifier stages. Thus for small input signals the RF amplifier will have high gain, but as the magnitude of the input signal increases, the gain of the RF amplifier decreases. This effect tends to keep the detector output relatively constant and prevents overdriving the RF into the saturation or cutoff regions. AGC is effectively and easily applied to a dual-gate MOSFET in the RF stage.

The Product Detector for SSB, DSB, and CW Demodulation

A product detector is a mixer used to down convert an input signal to baseband. The term, product detector, is normally used when referring to single sideband (SSB), or double sideband (DSB) demodulation, or continuous wave (CW). Essentially, it is a detector whose output is approximately equal to the product of the beat-frequency oscillator (BFO) and the RF signals applied to it. Output from the product detector is at audio frequency. Some RF filtering may be necessary at the detector output to prevent unwanted IF or BFO voltage from reaching the audio amplifier which follows the detector. In Figure 10, a product detector is used to demodulate a SSB signal. The lowpass filter that follows the mixer passes only the down conversion, or difference, frequency band. Since the SSB, or DSB, signal is transmitted with no carrier, or with a suppressed carrier, the frequency of the re-injected carrier in Figure 10 will not be exactly the same as that of the carrier that was suppressed in the generation of the SSB, or DSB, signal. If the frequency of the re-injected carrier is not sufficiently close, then the output audio will appear to have a "Donald Duck" quality as a result of all the demodulated frequencies being in error by a constant offset.

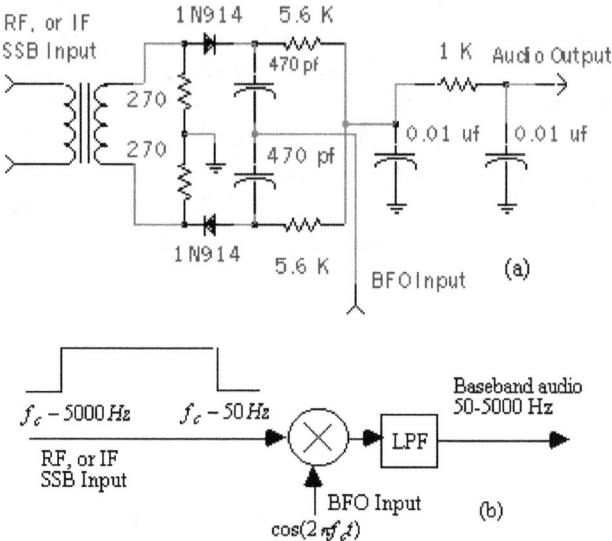

Figure 10: Product detector for demodulating a SSB, DSB, or CW signals, (a) schematic, (b) block diagram

The Quadrature Demodulator

In many modulation schemes, the carrier signal may contain both in-phase, and 90 degree quadrature phase components at the carrier frequency, wc. In the case of a simple product detector, a RF signal described by $\cos(\omega_c t + \varnothing)$ will produce a direct current (dc) baseband output signal when demodulated against $\cos(\omega_c t)$ if $\varnothing$ is $0°$, and will not produce any dc output for $\varnothing = 90°$ since

$$\left[\cos\left(\omega_c t + 90°\right)\right]\left[\cos\left(\omega_c t\right)\right] = 0$$

after filtering out the double frequency component. However, if the input signal is demodulated by both an in-phase, and a quadrature, local oscillator, then the baseband output signals, I and Q, can be measured to determine the amount of in-phase, and quadrature, energy in the input signal. The quadrature modulator, demodulator, will be used as a building block in subsequent sections for more elaborate modulation and demodulation schemes.

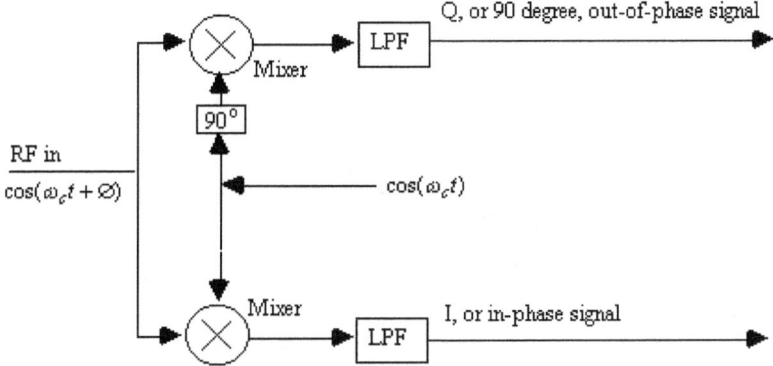

**Figure 11: Quadrature demodulator recovers both the in-phase and 90°
components of the input signal**

Costas Loop for Demodulating a BPSK Signal

A bipolar binary phase-shift-keyed signal (BPSK) is generated from a carrier signal, wc, that is modulated by shifting its phase by 0° or 180° at a specified baud rate. A simple mixer, or ordinary phase-locked loop cannot be used to recover the binary information since there are no spectral line components at ± wc. However, since the BPSK signal has a spectrum that is symmetrical with respect to the (suppressed) carrier frequency, a special type of phase-locked loop, called a Costas loop, may be used to supply the coherent reference signal for product detection. A Costas loop for BSPK demodulation is shown in Figure 12.

The Costas loop is analyzed by assuming that the VCO is locked to the input suppressed carrier frequency, wc, with a constant phase error of $\varnothing_e$. The bandwidth of the two lowpass filters are predetermined by the data rate. The two quadrature output signals are multiplied together and filtered with a lowpass that has a cutoff frequency near dc so that the filter acts as an integrator to produce the necessary dc control voltage, $K\sin(2\varnothing_e)$.

The Costas loop has a 180° phase ambiguity. Whenever the loop is energized, it is just as likely to phase lock so that the binary 1's come out as binary 0's, and vice versa. One of two methods can be used to resolve this $180°$ phase ambiguity. A known test signal could be sent over the system after the loop is turned on so that the sense of the polarity can be determined, or differential coding and decoding may be used.

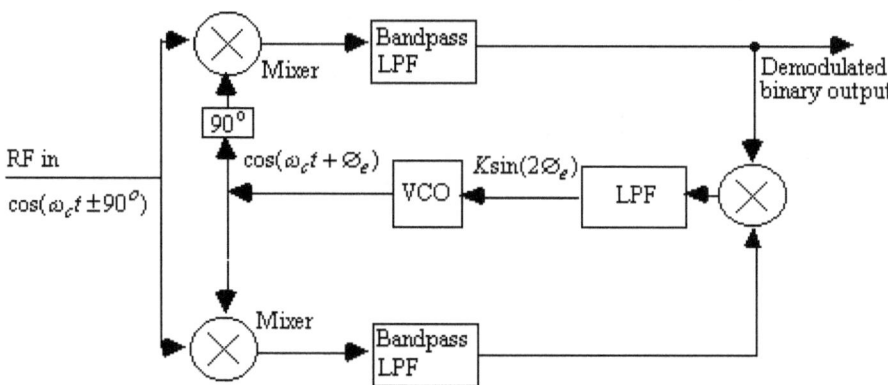

Figure 12: Costas loop for BPSK demodulation

Costas Loop for Demodulating a QPSK Signal

In quadrature phase-shift-keying (QPSK), the carrier signal can have one of four possible phases. Independent binary messages $x(t)$ and $y(t)$ modulate the two quadrature components of the carrier so that the transmitted signal can be represented as

$$V_{RFIN}(t) = x(t)\cos(\omega_c t + \emptyset_i) - y(t)\sin(\omega_c t + \emptyset_i)$$

After the usual multiplications and trigonometric identities, the low frequency output of the Costas loop will be

$$V_d(t) = [x(t)\sin(\emptyset_e) + y(t)\cos(\emptyset_e)]\text{sgn}[x(t)\cos(\emptyset_e) - y(t)\sin(\emptyset_e)]$$

$$-[x(t)\cos(\emptyset_e) - y(t)\sin(\emptyset_e)]\text{sgn}[x(t)\sin(\emptyset_e) + y(t)\cos(\emptyset_e)]$$

16

where $\varnothing_e = \varnothing_i - \varnothing_o$ and sgn( ) is the hard limiting operation. If the message waveforms are rectangular, the average dc outputs can be calculated as

$$V_d(t) = \sin(\varnothing_e) \qquad -45° < \varnothing_e < +45°$$

$$= -\cos(\varnothing_e) \qquad 45° < \varnothing_e < 135°$$

$$= -\sin(\varnothing_e) \qquad 135° < \varnothing_e < 225°$$

This characteristic is illustrated by the sawtooth in Figure 14.

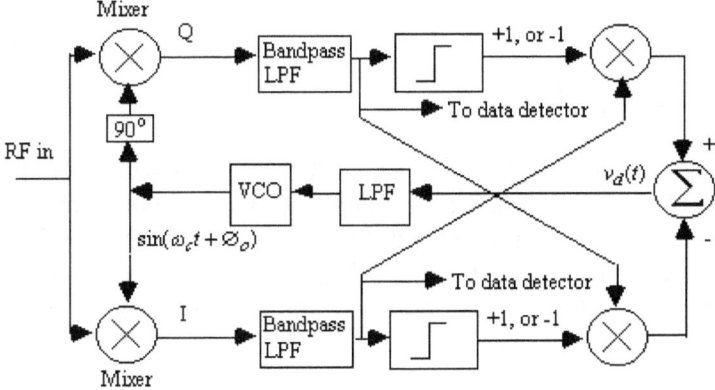

Figure 13: Costas loop for QPSK 4-phase demodulation

If the bandwidth is limited, the modulating pulses, x(t) and y(t), cannot be rectangular and the edges of the sawtooth in Figure 14 are rounded. Calculations are very difficult for rounded pulses since the pulse spectrums overlap in narrowband transmission. Noise will also cause rounding of the edges of the sawtooth.

A stable lock can be achieved at any of four different phases: 0, ± 90, or 180 °. There is an inherent fourfold ambiguity that must be resolved by other means.

The limiters shown in Figure 13 are essential for the correct operation of the circuit. If linear amplifiers were substituted for the limiters, the Costas loop would fail to lock.

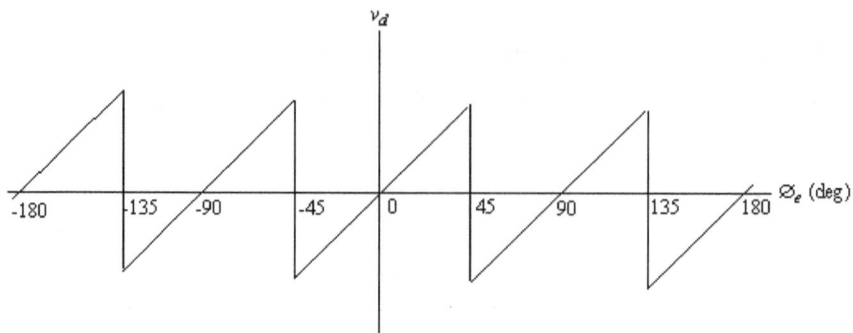

Figure 14: Phase detector characteristics of QPSK Costal loop

The Filter Method of SSB Generation

The double balanced modulator with two inputs at frequencies wc and wm, respectively, produces an output with frequencies wc + wm and wc - wm, neglecting carrier feedthrough, local oscillator leakage, and harmonic generation. Often it is required to reject either the sum, or the difference frequency, at the output. The filter method, phasing method, or the Weaver method can be used to eliminate the unwanted product.

Figure 15 shows the filter method in single sideband generation. The unwanted sideband is removed with a high Q crystal band pass filter. A major disadvantage to the filter method is that the filter is fixed-tuned to one frequency and must have an unrealistic sharp cutoff frequency to discriminate against the unwanted sideband. Assuming a 10 MHz carrier modulated by a 50 - 5000 Hz audio signal, a 13th order elliptic band pass filter is required with crystal Q's as high as 768,019 for 0.1 dB passband ripple, and 60 dB rejection of the unwanted sideband.

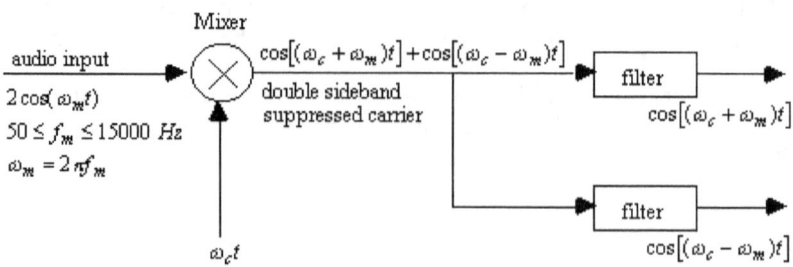

Figure 15: . Filter method of single sideband (SSB) generation

The Phasing Method of SSB Generation

The phasing method, shown in Figure 16, uses a Hilbert transformer to produce a constant 90 degree phase shift over a band of frequencies. The delay block matches the delay corresponding to the delay in the Hilbert transformer. In practice, a Hilbert transformer is approximated by a phase difference network in which the two outputs of the phase-difference network track each other over a band of frequencies with a constant $90°$ phase shift with unity gain at all frequencies.

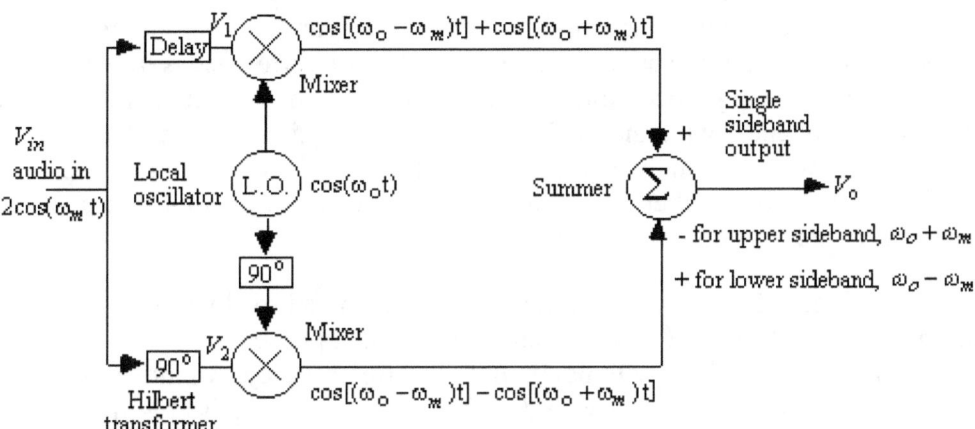

Figure 16: Phasing method of single sideband generation

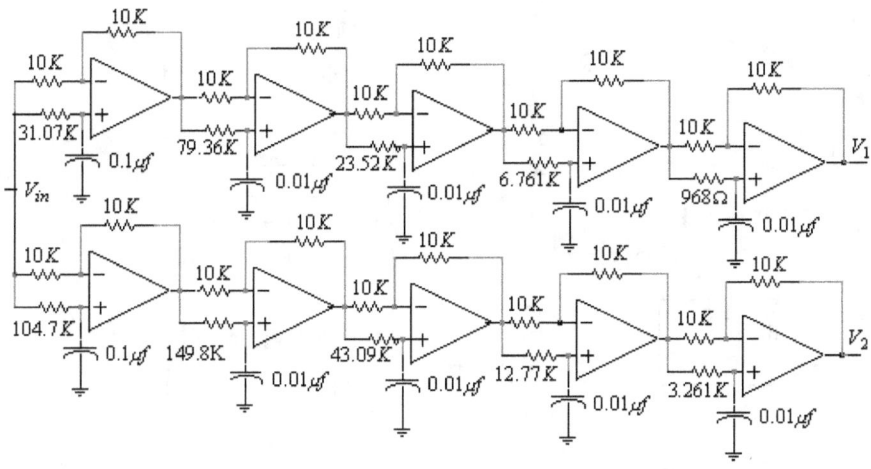

Figure 17: An active $90°$ phase splitter network

Figure 17 shows a phase difference network in which the two outputs are 90 ° apart over the frequency range of 50 - 5000 Hz, with a phase error of ±0.0607 °, resulting in a rejection of 65.5 dB for the unwanted sideband. In practice, such tolerances cannot be realized even with precision components and image rejecting mixers, based on this technique, usually do not realize more than 35 dB of rejection of the unwanted sideband.

The Weaver Method of SSB Generation

The Weaver method of single sideband generation in Figure 18 is similar to the phase method except that the audio broadband 90 ° phase difference is implement with a quadrature mixing processing. The frequency, w1, must be greater than the highest audio frequency, wm, and the lowpass filter is required to block the sum frequency, w1 + wm. Although an accurate broadband 90 ° phase difference is not required, the two lowpass filters must closely track each other in amplitude and phase for adequate rejection of the unwanted sideband.

The circuit in Figure 19 shows a technique for producing a broadband 90 ° (quadrature) phase shift for the local oscillator (L.O.). A phase locked loop (PLL) with an integrating loop filter will maintain 90 ° phase accuracy. An integrating PLL will maintain a quadrature phase relationship when the loop is closed, however PLL circuits tend to be noisy. Sideband noise is troublesome in both SSB and FM systems, but SSB is less sensitive to phase noise problems in the L.O.

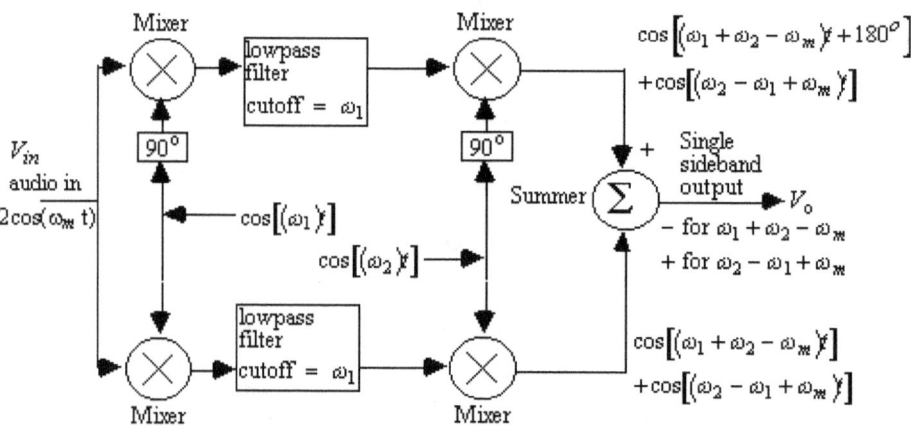

Figure 18: The Weaver method of single sideband generation

20

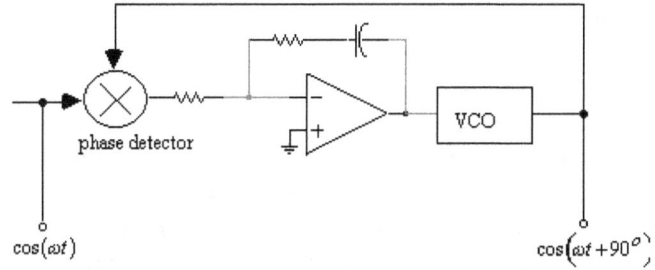

Figure 19: PLL quadrature synthesis

The Filter Method of Image Rejection

A typical AM super heterodyne radio uses an intermediate frequency (IF) of 455 Khz. If an AM station is transmitting on 1490 Khz, then the local oscillator must be tuned to 1490 + 455 Khz, or 1945 Khz, in order to convert the station frequency of 1490 Khz to the IF frequency of 455 Khz. However a station transmitting on 2400 Khz will also be converted to the IF frequency of 455 Khz unless a preselection filter is used to prevent the 2400 Khz signal from reaching the input of the mixer. The 2400 Khz station is referred to as the image.

Traditionally, a tuned filter is used at the input of the mixer to reject the image frequency (Figure 20). In AM and FM radios, the front end tuned filter is designed to track the local oscillator (L.O.) by a constant difference of 455 Khz (or 10.7 MHz in FM radios). This is usually done by a specially designed double ganged variable capacitor that sets both the frequency of the L.O. and the passband frequency of the front end filter simultaneously.

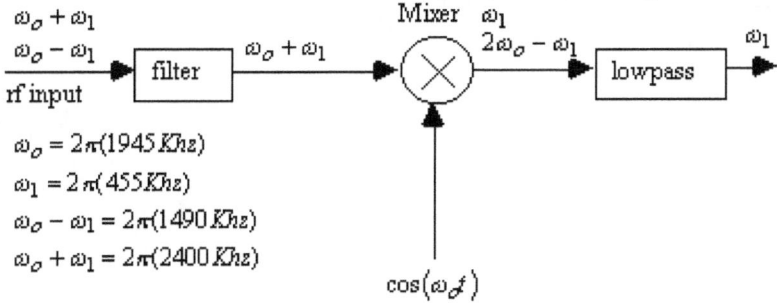

$$\omega_o = 2\pi(1945\,Khz)$$
$$\omega_1 = 2\pi(455\,Khz)$$
$$\omega_o - \omega_1 = 2\pi(1490\,Khz)$$
$$\omega_o + \omega_1 = 2\pi(2400\,Khz)$$

Figure 20: Filter method of image rejection.

The Phasing Method of Image Rejection

In the analysis of the phasing method in Figure 16, V1 consists of a $\cos(w_1t)$ term from the upper image and a $\cos(w_1t)$ from the lower image. V2 consists of a $\cos[(w_1t)-90°]$ term from the upper image and a $\cos[(-w_1t)-90°]$ term from the lower image. A property of the Hilbert transform is that it shifts all positive frequencies by $+90°$ and all negative frequencies by $-90°$. Therefore, the two terms, after the Hilbert transform, become $\cos(w_1t)$ from the upper image, and $\cos[(-w_1t)-180°]$ for the lower image. Since $\cos(A) = \cos(-A)$, the $\cos[(-w_1t)-180°]$ term is the same as $\cos[(w_1t)+180°]$. Adding, or subtracting, in the summer can then select whether the IF frequency, w1, is derived from the upper, or the lower, RF image.

The $90°$ phase difference network of Figure 17 could be used to realize the Hilbert transform in Figure 16, by separating the inputs to the two sections in Figure 16, so that V1 and V2 are not shorted together. Figure 22 shows an alternate implementation of a $90°$ phase difference network based primarily on resistors and capacitors. The ARRL handbook (1) recommends all resistors 12K, with C1 = 0.044 μF, C2 = 0.033 μF, C3 = 0.02 μF, C4 = 0.01 μF, C5 = 5600 pF, and C6 = 4700 pF for 60 dB of sideband suppression over the range of 300 to 3000 Hz. It is important that all resistors and capacitors be closely matched for good performance.

The values for RiCi for Figure 22 are calculated using elliptic functions. Typically the worst case suppression is desired to be the highest possible. This requires an equal ripple, or Tcheybscheff approximation. The mathematics are straightforward and given in detail by Saraga(2). For an upper and lower frequency of fu and fl respectively, the RiCi values for an n-section network are

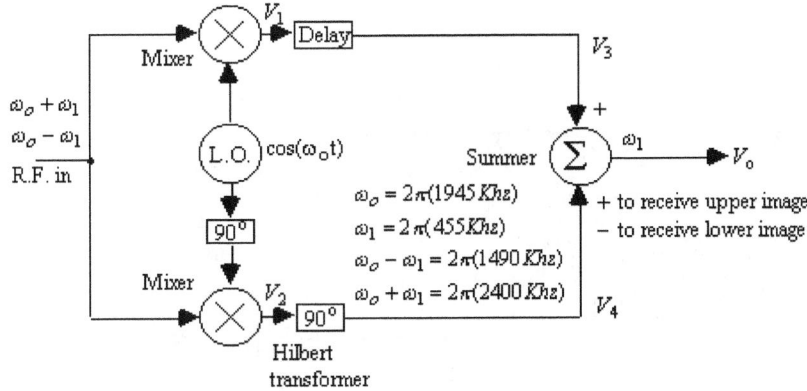

Figure 21: Phasing method of image rejection

$$R_iC_i = \frac{dn\left[\dfrac{2i-1}{2n}\right]K(k),k}{2\pi f_L}$$

where

$$k = \sqrt{1-\left(\frac{F_L}{F_u}\right)^2} \ ,K(k)$$

is the complete elliptic integral of the first kind, and dn(u,k) is a Jacobi elliptic function with R1-C1 defining section 1, and so on. Based on the previous equations, table 2 gives the optimal Tchebyscheff values for the R-C phasing network, fL and fu are the lower and upper frequencies, n is the number of sections, and fi, where i is 1 through n, are the frequencies of exact $90\,^\circ$ phase shift. The corresponding R-C values are $1/(2\pi fi)$. Sup-dB is the minimum sideband suppression over the network range in dB. The network sections are ordered from the largest RC value to the smallest.

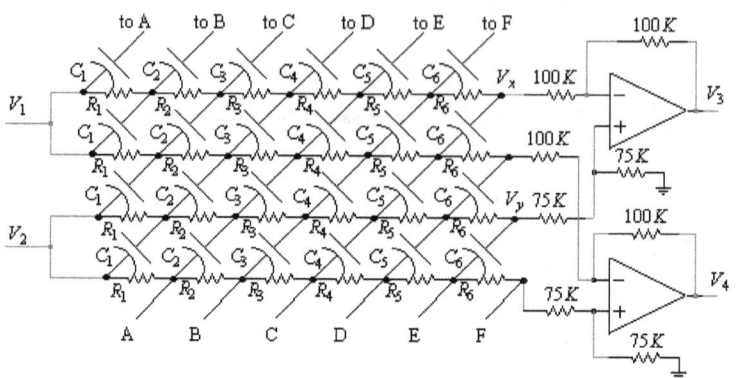

Figure 22: Six section RC 90° phase difference network

Table 2

R-C values for 90° phase shift network

fL	fu	n	Sup-dB	f1	f2	f3	f4	f5	f6	f7	f8
300	3000	4	40.5	332.2	629.8	1429.0	2709.0				
300	3000	5	52.1	320.5	500.7	948.7	1797.6	2808.1			
300	3000	6	63.7	314.2	435.5	720.3	1249.5	2066.8	2864.5		
300	3000	7	75.4	310.4	397.8	595.3	948.7	1511.8	2262.4	2899.4	
300	3000	8	87.0	308.0	374.0	519.4	771.2	1167.0	1732.7	2406.2	2922.5
200	4000	5	42.9	219.5	398.4	894.4	2008.1	3645.0			
200	4000	6	52.7	213.5	332.1	633.1	1263.6	2408.9	3747.8		
200	4000	7	62.5	209.9	294.6	497.5	894.4	1608.2	2715.5	3812.0	
200	4000	8	72.2	207.5	271.2	417.8	689.9	1159.6	1915.0	2949.6	3854.8
150	6000	6	44.7	163.6	287.7	628.9	1431.1	3128.3	5500.9		
150	6000	7	53.1	160.0	247.7	471.0	948.7	1910.7	3633.0	5626.4	
150	6000	8	61.5	157.6	223.1	381.3	696.7	1291.9	2360.2	4033.2	5710.4

One last comment is needed on the design of the operational amplifier circuit shown in Figure 22. The input impedance at Vy is 150K. The current drawn toward the inverting input from Vx is

$$\frac{V_x - \frac{V_y}{2}}{100K}$$

and since Vx = -Vy with perfect phasing, the impedance seen at Vx in the direction of the operational amplifier is 150K. Making all four resistors of equal value would reduce the sideband suppression to about 35 dB.

The Weaver Method of Image Rejection

The Weaver method of image rejection is summarized in Figure 23.

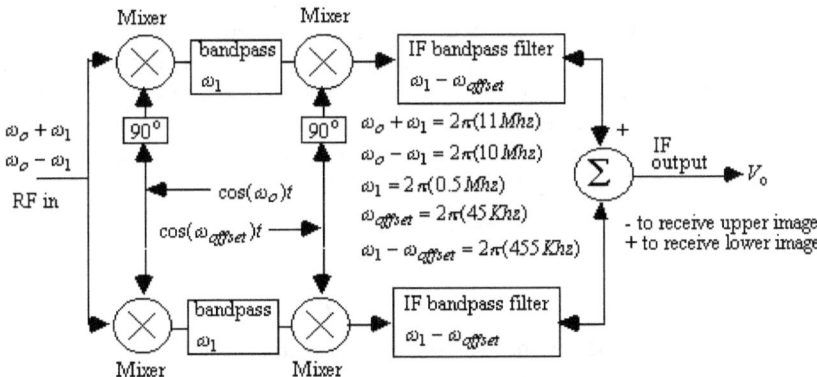

Figure 23: The Weaver method of image rejection

CHAPTER 3

Gilbert Cells

A Gilbert cell is a cross-coupled differential amplifier, similar to the topology in Figure 24, where the gain is controlled by modulating the emitter bias current. The amplitude of a differential input RF signal, applied to pins 6 and 7 of the HFA3101, can be linearly controlled by a differential ac voltage applied to pins 1 and 4. Because the gain control is highly linear, Gilbert cells are often referred to as four-quadrant multipliers and have common applications as mixers, AGC amplifiers, amplitude modulators, double sideband (DSB) modulators, single sideband (SSB) modulators, AM detectors, SSB and DSB detectors, frequency doublers, squaring circuits, dividers, square-root circuits, and root-mean-square, rms., measuring circuits. In order to understand how the Gilbert cell operates, it is necessary to review some fundamental concepts of bipolar transistors.

Bipolar Junction Transistor Models

Figure 25 shows two equivalent small signal models for a bipolar transistor. All ac components will be represented by a lower case letter with *vbe* denoting the ac input voltage across the base-emitter junction of the transistor, and *ic* will be the corresponding ac collector current. The transconductance, *gm*, of a transistor is set by its dc quiescent collector current. In Figure 26, the dc quiescent collector current is denoted by Io/2, therefore

$$g_m = \frac{I_o/2}{V_T} \tag{1}$$

where VT is the thermal voltage and is taken at 0.025 volts at room temperature.

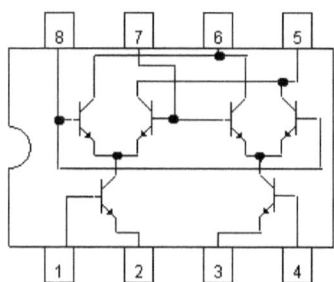

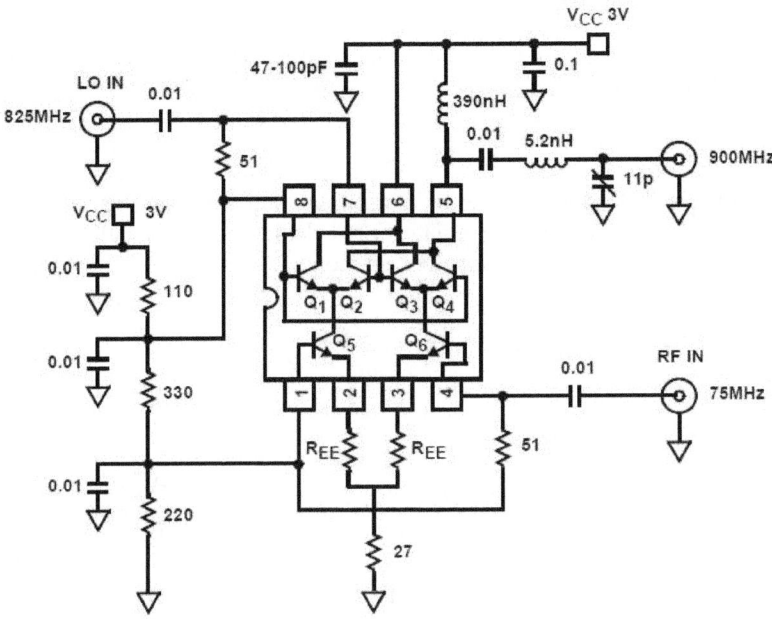

Figure 24: Harris HFA3101 5 GHz Gilbert cell array with application

Once *gm* is established by setting the value of the dc quiescent collector current, Io/2, the gain can be derived from

$$gm\ vbe = ic = \beta\ ib \tag{2}$$

where β is the current amplification of the transistor and is dependent upon the transistor selected.

The input impedance for an ac signal between the base-emitter junction of a transistor will be denoted by r_π and is given by

$$r_\pi = \frac{\beta+1}{g_m} \tag{3}$$

Equations (1-9) are small signal approximations and they are valid for *vbe* less than 10 millivolts.

The Differential Amplifier

Figure 26 shows a differential amplifier using a constant current source. The following equations immediately provide the means to analyze Figure 26. The ac voltage across the base-emitter junction of Q1 is

$$, \quad v_{be} = \frac{r_\pi v_{in}}{2(r_\pi + R_{in})} \tag{4}$$

or

$$v_{out1} = -i_c R_c = -g_m v_{be} R_c = -g_m \frac{r_\pi v_{in} R_c}{2(r_\pi + R_{in})} \tag{5}$$

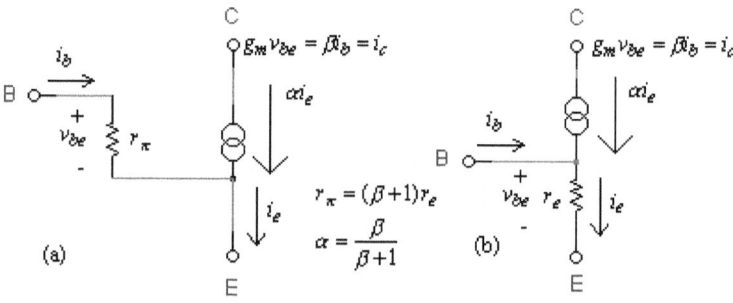

Figure 25: Equivalent small signal models for a bipolar transistor.

The ac gain from *vin* to *vout1* is then

$$\frac{v_{out1}}{v_{in}} = -g_m \frac{r_\pi R_c}{2(r_\pi + R_{in})} \tag{6}$$

Notice here that if Rin $\ll$ than r_π, the total gain is linearly dependent on the value of the constant current source, Io. In this case

$$\frac{v_{out1}}{v_{in}} = -g_m \frac{R_C}{2} \tag{7}$$

The total voltage at *vout1* is

$$v_{out1} = -g_m \frac{R_C}{2} v_{in} + V_{cc} - R_C \left(\frac{I_o}{2} \right) \tag{8}$$

Similarly, at the collector of Q2, the output voltage, *vout2* is

$$v_{out2} = g_m \frac{R_C}{2} v_{in} + V_{cc} - R_C \left(\frac{I_o}{2} \right) \tag{9}$$

with the assumption that both transistors are well matched.

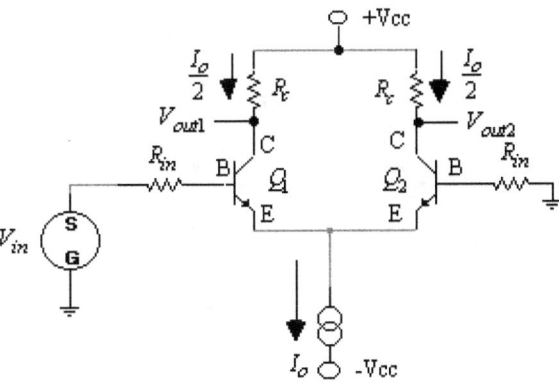

Figure 26: Differential amplifier with constant current source.

The Single Balanced Modulator

Figure 27, the concept of a differential amplifier is now extended to a single balanced modulator by modulating the constant current source with a low frequency signal such that

$$I = Io + k1 \cos(wmt) \text{ for } k1 < Io. \tag{10}$$

If vin is also represented as a sine wave such that

$$vin = k2 \cos(wct) \tag{11}$$

then

substituting equations (10) and (11) into equations (8) and (9) provide

$$v_{out1} = -\left[\frac{I_o + k_1 \cos(\omega_m t)}{2V_T}\right]\left(\frac{R_c}{2}\right)k_2 \cos(\omega_c t) + V_{cc} - R_c \frac{I_o + k_1 \cos(\omega_m t)}{2} \tag{12}$$

and

$$v_{out1} = +\left[\frac{I_o + k_1 \cos(\omega_m t)}{2V_T}\right]\left(\frac{R_c}{2}\right)k_2 \cos(\omega_c t) + V_{cc} - R_c \frac{I_o + k_1 \cos(\omega_m t)}{2} \tag{13}$$

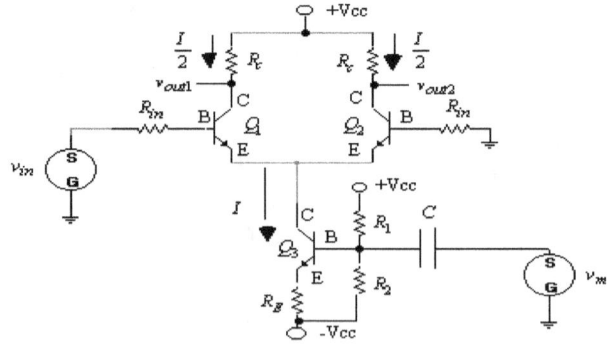

Figure 27: Single balanced modulator

It is seen that both (12) and (13) contain frequency components at four distinct frequencies, including wc, wc - wm, wc + wm, and wm. By taking the difference of vout1 and vout2 with a difference amplifier, the baseband term at frequency, wm, is eliminated leaving

$$v_{out} = \left[\frac{I_o + k_1 \cos(\omega_m t)}{V_T}\right]\left(\frac{R_c}{2}\right) k_2 \cos(\omega_c t)$$

(14)

or, by using a trigonometric identify,

$$v_{out} = \left[\frac{I_o R_c k_2}{2V_T}\right]\cos(\omega_c t) + \left(\frac{I_o R_c k_2}{4V_T}\right)\left([\cos(\omega_c + \omega_m)t] + [\cos(\omega_c - \omega_m)t]\right)$$

(15)

The balanced modulator with the difference amplifier that implements equation (15) is shown in Figure 28. The corresponding voltages as seen on an oscilloscope for vout1, vout2, and vout are shown in Figure 29.

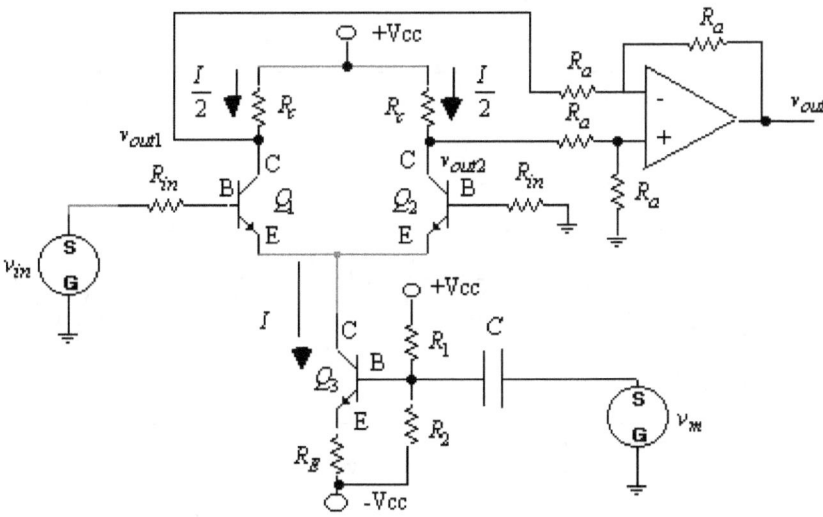

Figure 28: Single balanced modulator with difference amplifier

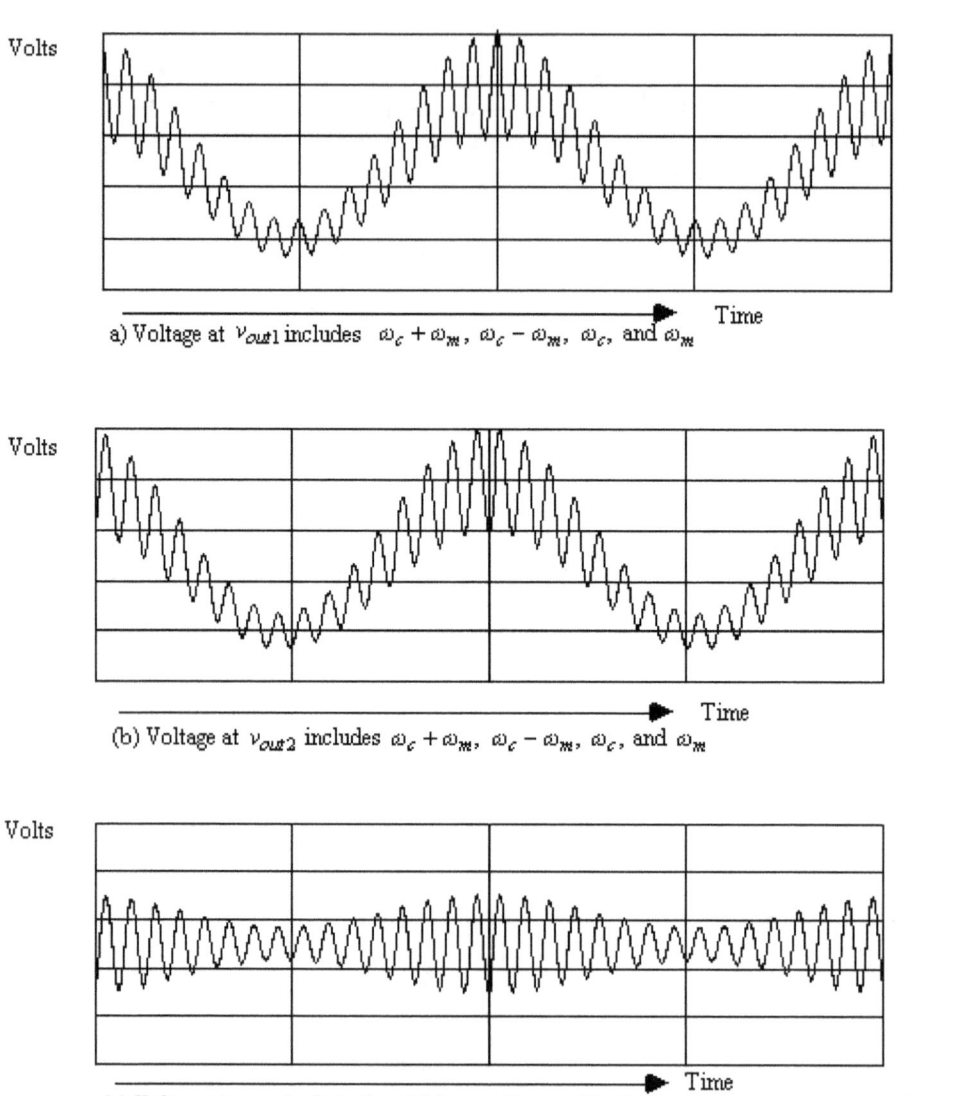

Volts

a) Voltage at v_{out1} includes $\omega_c + \omega_m$, $\omega_c - \omega_m$, ω_c, and ω_m

Volts

(b) Voltage at v_{out2} includes $\omega_c + \omega_m$, $\omega_c - \omega_m$, ω_c, and ω_m

Volts

(c) Voltage at v_{out} is that of an AM waveform and includes $\omega_c + \omega_m$, $\omega_c - \omega_m$, and ω_c

Figure 29: Voltages at *vout1*, *vout2*, and *vout*, for a single balanced modulator with difference amplifier

The Double Balanced Modulator

The double balanced modulator eliminates the carrier frequency at wc and effectively implements a mixer that generates only the sum and difference frequencies. An extension of the balanced modulator to create a double balanced modulator is shown in Figure 30. By analogy with the balanced modulator, using equations (12) and (13), the equations for the double balanced modulator in Figure 30 become

$$v_{out} = \left[\frac{I_o + k_1\cos(\omega_m t)}{2V_T}\right]\left(\frac{R_c}{2}\right)k_2\cos(\omega_c t) + V_{cc} - R_c\frac{I_o + k_1\cos(\omega_m t)}{4}$$

$$-\left[\frac{I_o - k_1\cos(\omega_m t)}{2V_T}\right]\left(\frac{R_c}{2}\right)(-1)k_2\cos(\omega_c t) + V_{cc} - R_c\frac{I_o - k_1\cos(\omega_m t)}{4}$$

$$= -\left[\frac{k_1 k_2 R_c}{4V_T}\right]\left[\cos(\omega_c + \omega_m)t + \cos(\omega_c - \omega_m)t\right] + V_{cc} - R_c\frac{I_o}{2} \tag{16}$$

and

$$v_{out2} = \left(\frac{k_1 k_2 R_c}{2V_T}\right)\left[\cos(\omega_c + \omega_m)t + \cos(\omega_c - \omega_m)t\right] + V_{cc} - R_c\frac{I_o}{2} \tag{17}$$

The difference amplifier at the output of Figure 30 eliminates the dc term leaving

$$v_{out} = \left(\frac{k_1 k_2 R_c}{2V_T}\right)\left[\cos(\omega_c + \omega_m)t + \cos(\omega_c - \omega_m)t\right] \tag{18}$$

The output waveform, vout, for the double balanced modulator is shown in Figure 31.

The double balanced mixer in figure 30 generates both a sum and difference frequency. Further derivatives of the double balanced mixer can be used to generate only the upper, or lower, sideband during the modulation process, or can be used to mix two signal and reject either the lower, or the upper, image frequency. The analysis details are based on the trigonometric identity

$$\cos(a)\cos(b) = \frac{1}{2}\cos(a-b) + \frac{1}{2}\cos(a+b) \tag{19}$$

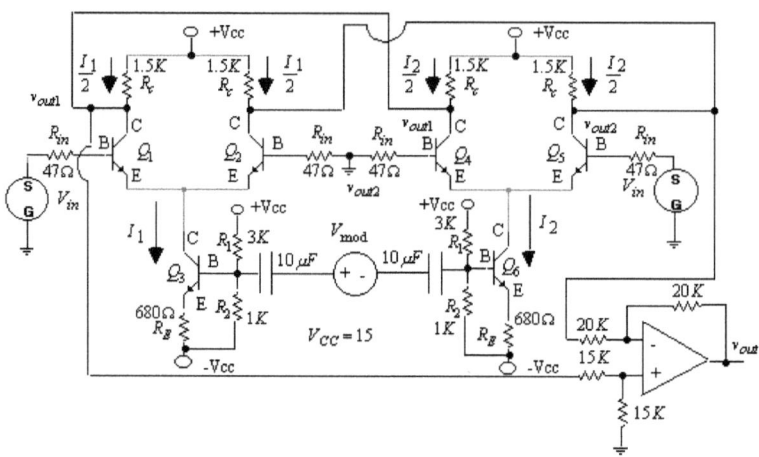

Figure 30: Double balanced modulator with elimination of the dc component at *vout*

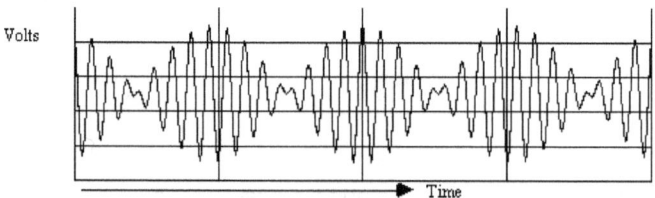

Figure 31: Typical output waveform for a double balanced modulator shows a double sideband (DSB) waveform and includes only the sum and difference frequencies, $\omega_c + \omega_m$ and $\omega_c - \omega_m$.

CHAPTER 4 - Types of Mixers

The Pentagrid Mixer

The pentagrid mixer is shown here for historical reasons and uses a specially designed vacuum tube to both oscillate (local oscillator or LO) and to amplify the incoming signal, RFin. It also mixers the locally generated signal with the receive signal to produce four frequencies, which includes the RFin, the LO, the RFin-LO, and the RFin+LO frequencies. The desired intermediate frequency in an AM radio is usually the RFin-LO which is 455KHz (the intermediate frequency) and filtered to eliminate the other three frequencies. The converter stage uses a 6A8 pentagrid tube. Part of the tube functions as a tuned amplifier and the other section functions as the local oscillator. Grids G5, G4, and G3, together with the plate and cathode form a tetrode amplifier. Grids G1 and G2 together with the cathode form a triode oscillator.

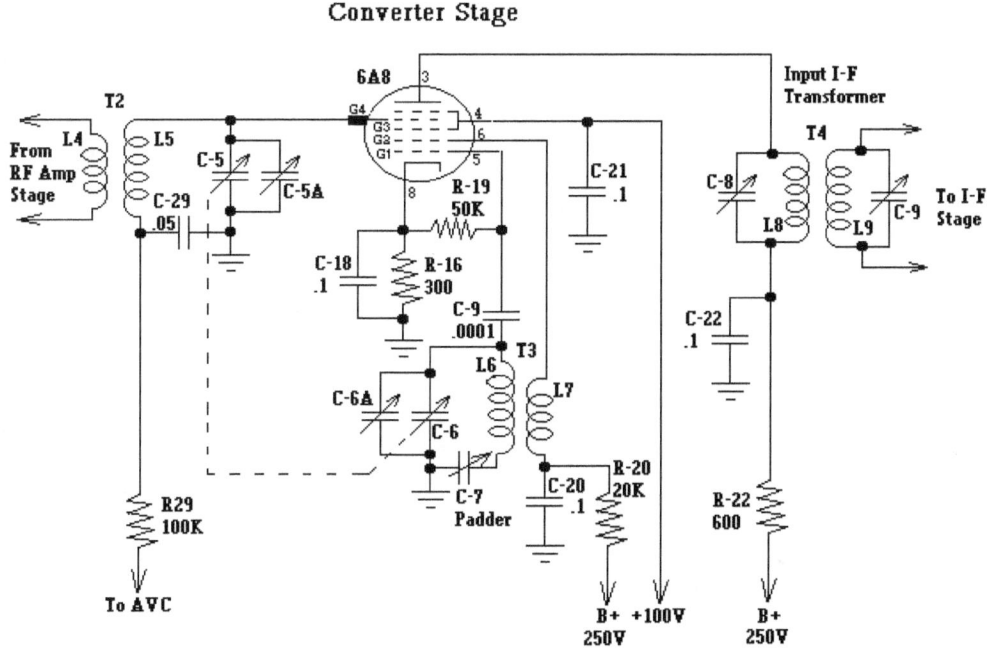

Figure 32: Pentagrid Mixer

The input to the converter stage is RF transformer T2 Tuning of the input is accomplished by the circuit consisting of L5, and C-5 tunes the input and feeds the signal to grid G4. The amplified signal will appear at pin 3 of the 6A8. Capacitor C-5 is one section of the tuning capacitor, ganged to C-6 of the oscillator stage. The oscillator section uses the cathode, grid G1, and grid G2 of the 6A8, and oscillator transformer T3. Grid G1 is the oscillator grid, while grid G2 is effectively the anode or plate of the oscillator. Feedback to maintain oscillation is done by the coupling of coils L6 and L7. Tuning capacitor C-6 across L-6 tunes the oscillator frequency. The C5-C6 capacitor is dual ganged capacitor of the cut-plate variety as mentioned in chapter 1.

Bipolar transistor mixer.

As transistors came into common use in the 1950s, the initial urge was to replace each vacuum tube with a single transistor that does the same function. In the following schematic, the SFT308 is both an RF amplifier and local oscillator and uses the non-linear region of the SFT308 to generate sum and difference frequencies. SFT308 section uses a cut-plate capacitor to maintain the local oscillator at a frequency of 455Khz higher than the frequency of the incoming signal.

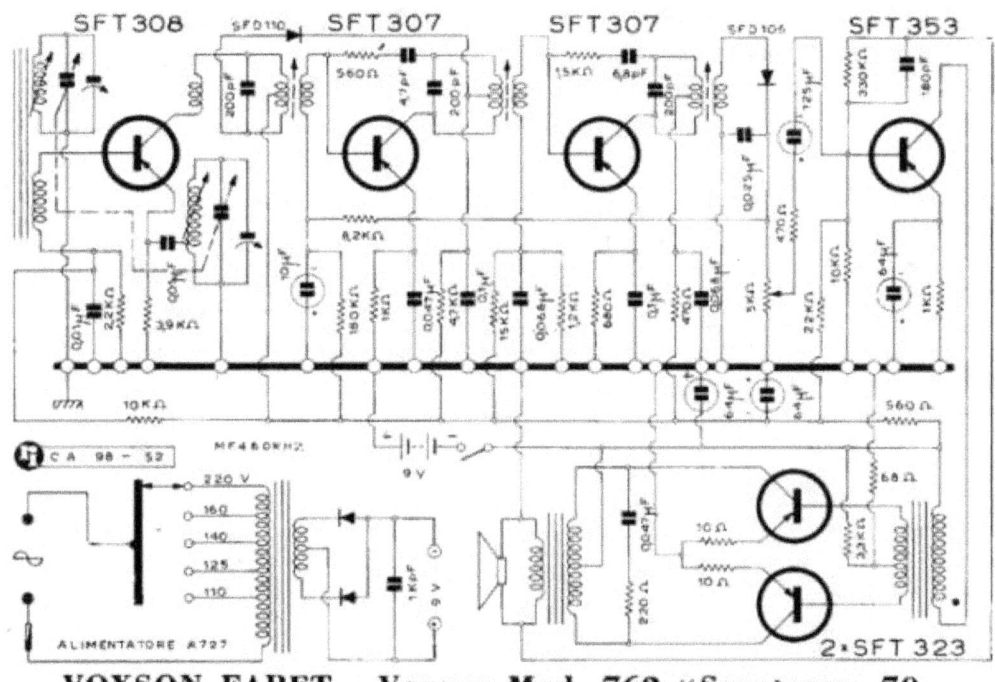

VOXSON FARET - Voxson Mod. 762 "Sportsman 70,,

Figure 33: Bipolar transistor mixer in a 6 transistor radio

Dual Gate MOSFET Mixer

The schematic below is a broadband mixer that uses a separate local oscillator. Y1 selects the LO frequency. Since the RF input is not tuned, the intermediate frequency amplifier can actually be a separate AM radio that tunes through the selected band. If Y1 is a 4.5MHz LO, and if it is desired to tune though 2-3MHz, then the difference frequency would range between 6 MHz to 7 MHz, depending on the frequency of the RF signal (to Ant). A frequency of 6-7MHz would normally create an intermediate frequency of 0.5MHz to 1.5MHz, if not for the lowpass filter at the gate of Q1 which is designed to block any RF frequency above 4.5MHz, so that the image frequency of 6-7MHz cannot pass into the dual gate MOSFET.

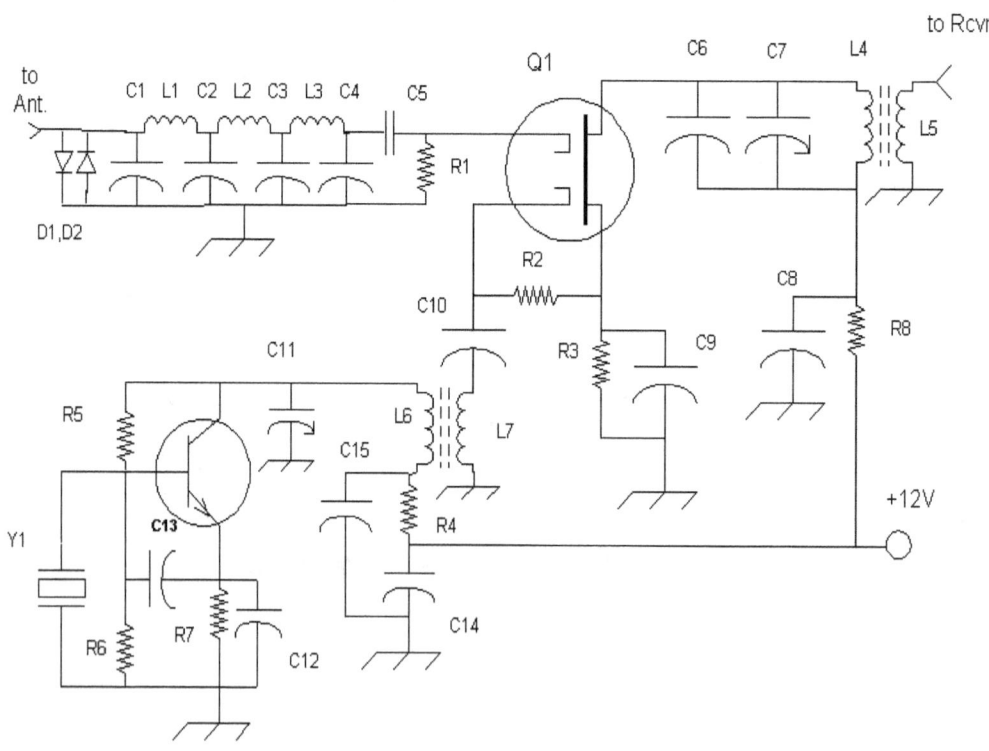

Figure 34: Dual Gate MOSFET Mixer

Another circuit using MESFET technology is shown below.

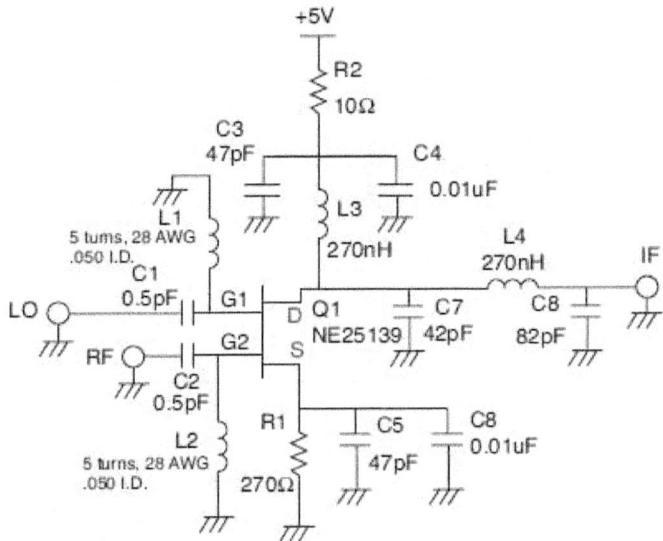

Dual-Gate MESFET Active Mixer

Figure 35: Dual Gate MESFET Mixer

S-Parameter	70 MHz(IF)	845 MHz(LO)	915 MHz(RF)
S_{11}	0.99 < -3°	0.95 < -25°	0.94 < -29°
S_{21}	2.36 < 177°	2.23 < 142°	2.45 < 137°
S_{12}	0.001 < 87°	0.005 < 79°	0.005 < 79°
S_{22}	0.97 < -1°	0.96 < -9°	0.99 < -13°

The doubly balanced diode mixer

A diode double-balanced mixer is the most used mixer in that it consists of passive components that can work up to microwave frequencies. This mixer contains a diode ring consisting of 4 diodes and two unbalanced-to-balanced transformers. Each leg consists of up to four diodes. Input and output ports-are named local oscillator (LO), radio frequency (RF), and intermediate frequency (IF)-connect the doubly balanced mixer to its associated circuitry. The manufacturer will provide a recommended input level for the (LO) and an input and output impedance to match the IF and the LO. Usually these mixers are available in 50 ohm, or 75 ohm, versions.

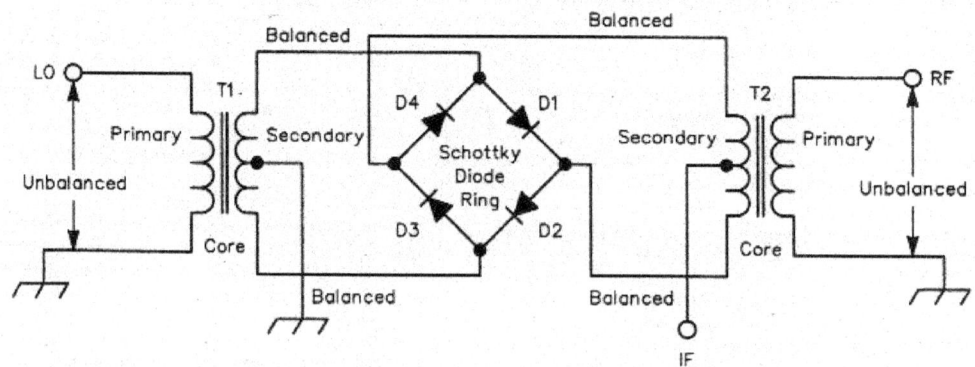

Figure 36: Doubly Balanced Diode Mixer

The doubly balanced MOSFET mixer.

This diagram shows the basic concept of a double balanced FET mixer. Double balanced FET mixers using discrete components can sometimes be optimized to provide better performance figures.

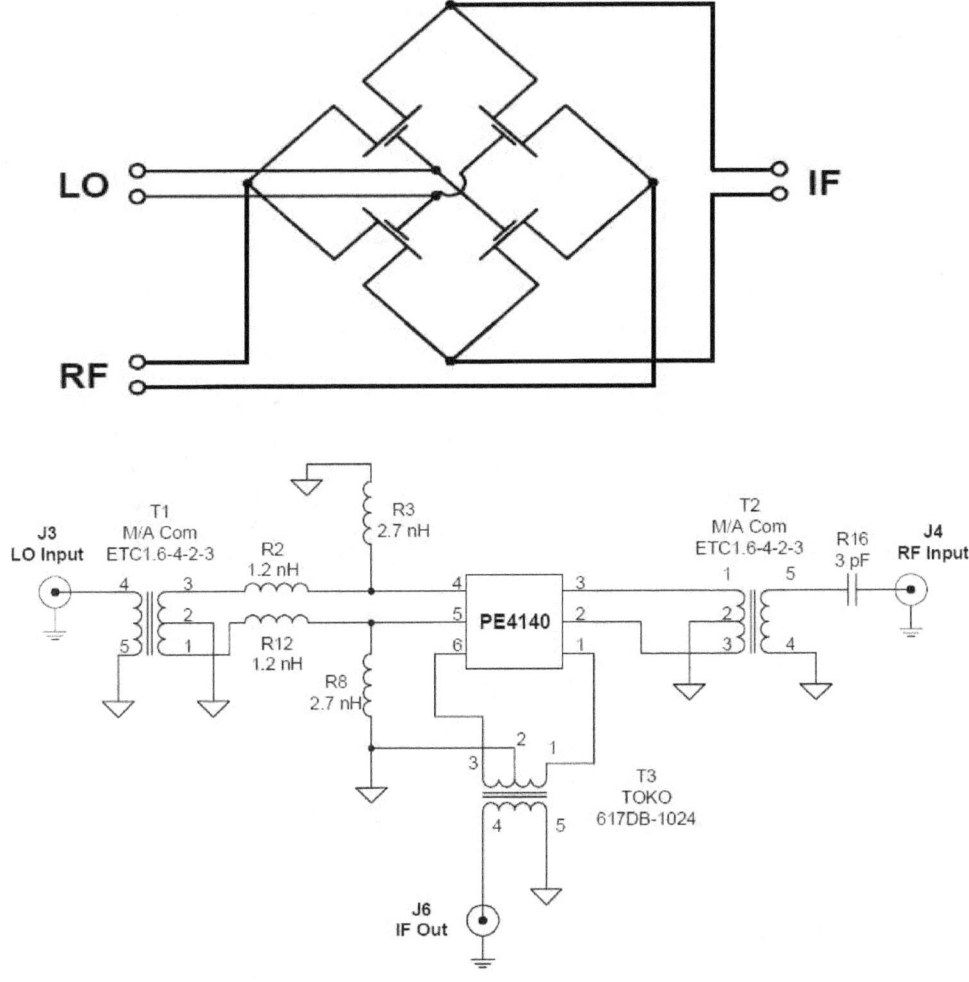

Figure 37: Doubly Balanced MOSFET Mixer

The Triple Balanced Mixer.

Double Balanced Mixers are most often used for frequency translation. The advantages of the double balanced mixer are good local oscillator to RF and local oscillator to IF isolation with some RF-IF isolation. Triple Balanced Mixers use three separate baluns for RF, LO and IF ports. This helps provide wider bandwidth with improved L-R, L-I and R-I isolations, return loss, compression and IP3, but at the expense of 3dB higher LO power.

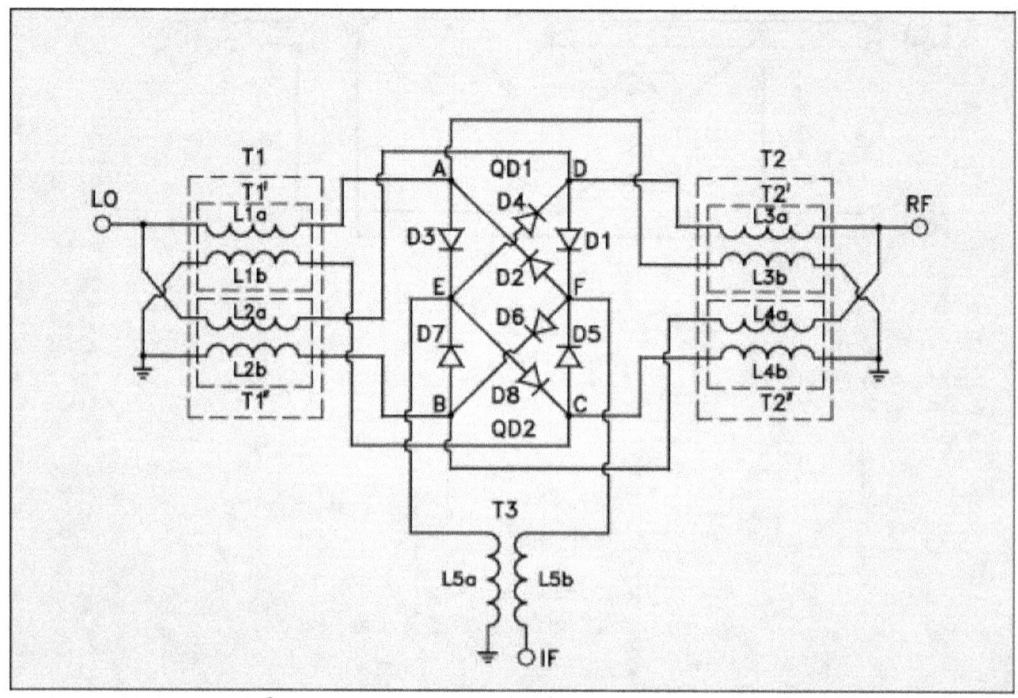

Compliments of Mini-Circuits

Figure 38: The Triple Balanced Diode Mixer

Mixers in Phase Lock Loops.

The block diagram below shows an analog phase-locked loop (PLL) circuit. The VCO produces a sine wave on its output . The frequency of the sine wave is proportional to the input voltage to the VCO (Vin)

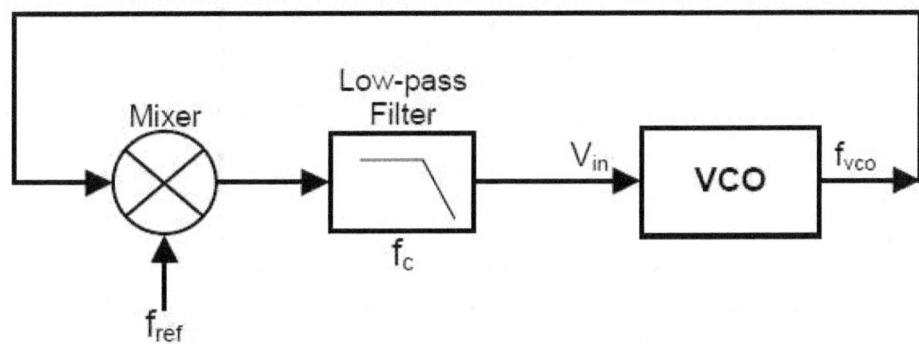

Figure 39: Phase Locked Loop with Mixer

Below is a block diagram of a digital PLL (DPLL). The key differences between a digital PLL and an analog PLL are:
1. The waveform used for the reference signal as well as the waveform at the output of the VCO are square waves (instead of sinusoids as in the analog PLL)
2. An exclusive-OR (XOR) gate can be used for the phase detector (instead of a mixer as in the analog PLL)

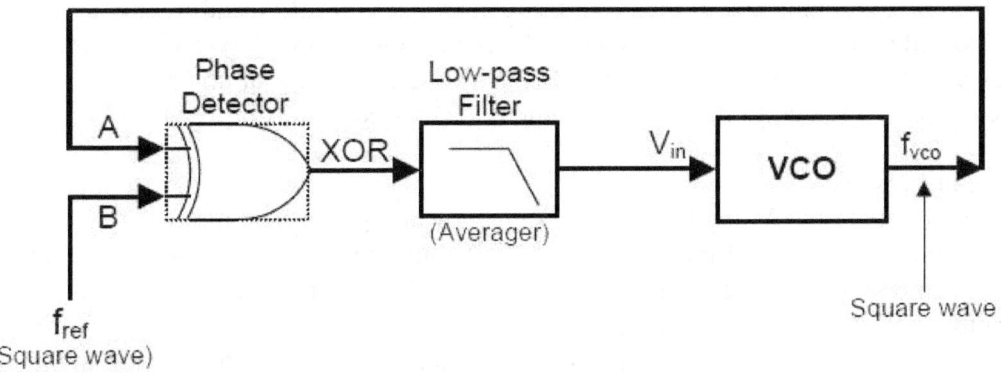

Figure 40: Phase Locked Loop using Digital XOR Gate as a Mixer

The phase-frequency detector is more commonly used as opposed to a mixer, or XOR, gate is a digital PLL. The circuit diagram is shown below.

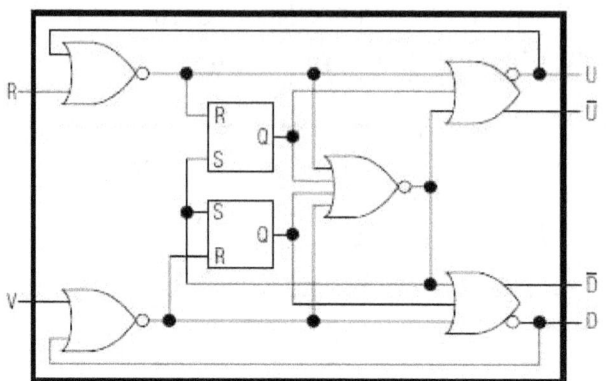

Figure 41: Phase Lock Loop using Phase-Frequency Detector as a Mixer

The phase-frequency detector compares a VCO input, V, to a reference input, R, to determine the phase or the frequency difference between the V and the R inputs. The differential outputs U, U* and D, D* are output pulse trains with aduty cycle proportional to the frequency or phase difference between the R and V inputs.

These outputs are the high and low signals needed to control the PLL VCO. Subtracting and integrating the U and D outputs provide the VCO control signal. The detector can detect phase differences up to ±2π.

Frequency Detection
When the two inputs are at the same frequency, but input R is leading input V, the device alternates between states 0 and 2. Also, if the R input lags the V input, the device alternates between states 0 and 1. With the two inputs at unequal frequencies, the output becomes dependent of the frequency difference.

Output Pulses
When inputs R and V are at the same frequency and phase, outputs U, U* and D, D* produce a stream of minimum duty cycle pulses that occur at the same time as rising edges of the input waveforms. The PLL is locked. If either input (R or V) begins to lead the other in phase, the width of pulses on the corresponding output (U output for R input, D output for V input) increases in direct proportion to the phase difference. In a phase locked loop implementation, these outputs direct the system VCO to increase or decrease frequency to maintain the lock condition.

CHAPTER 5 – Math Operations

The simplest electronic multipliers (mixers) is to use logarithmic amplifiers. The computation uses the fact that the antilog of the sum of the logarithm of two numbers is the product of those numbers.

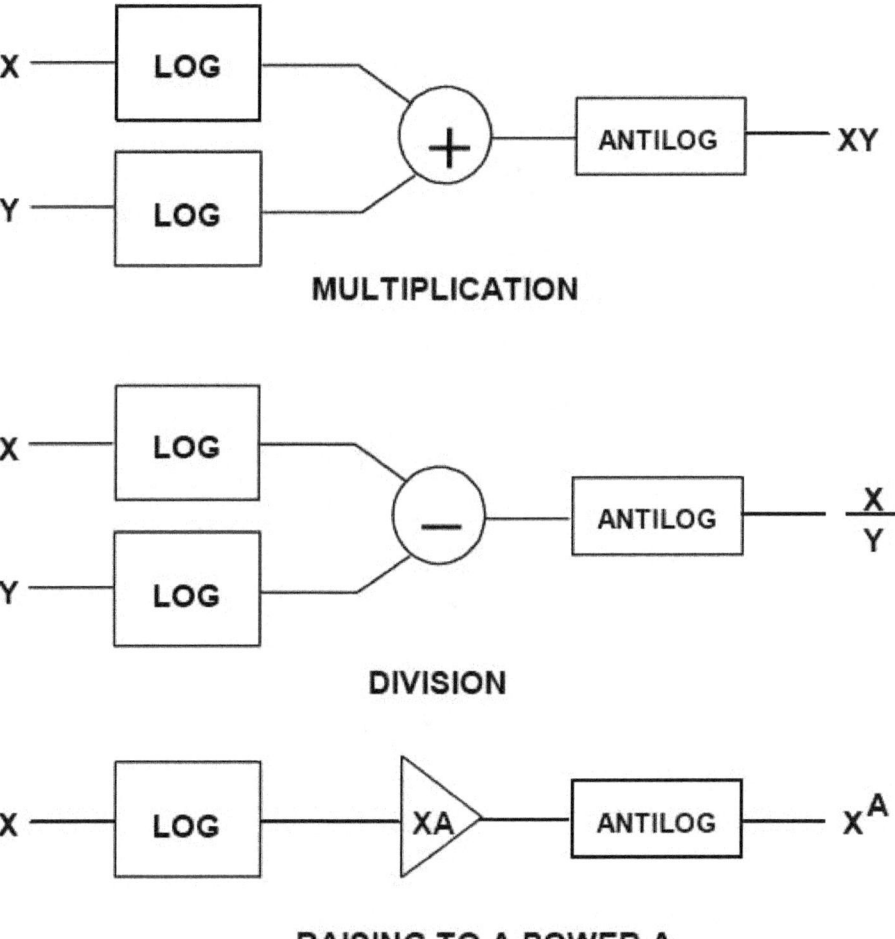

Figure 42: Using Log Amps to Implement Mixer Mathematical Functions

Division and Square Root Operations

Division and square-rooting are done by placing the multiplier (mixer) in the feedback path of an operational amplifier. Since most multipliers use an operational amplifier, an external jumper permits the same circuit to perform in either mode.

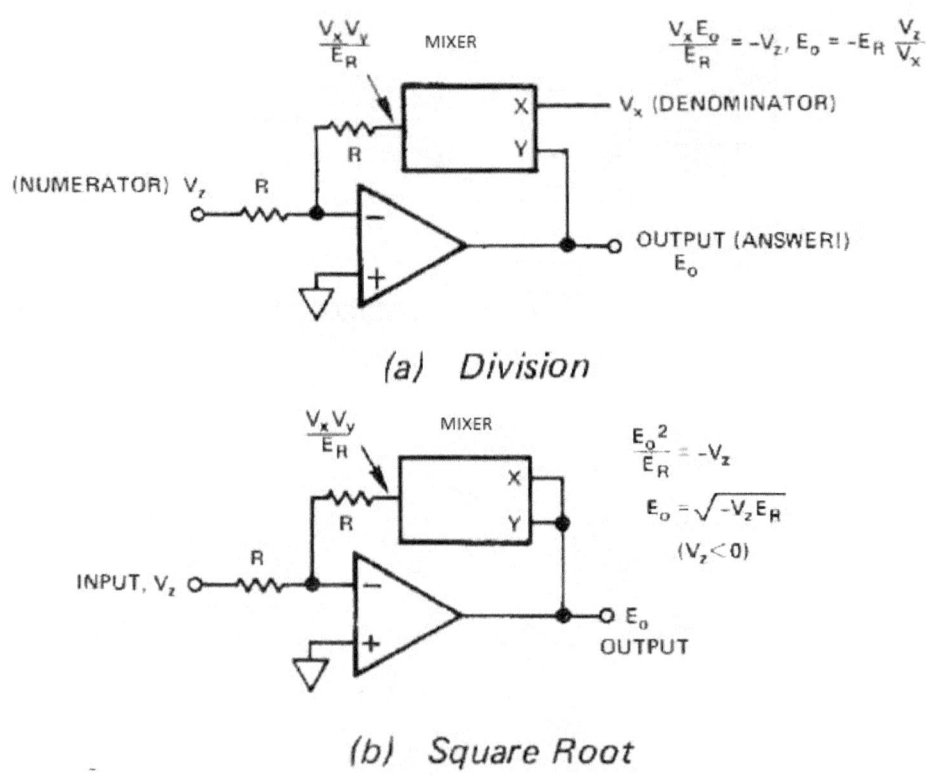

(a) Division

(b) Square Root

Figure 43: Division and Square Root Functions using Mixers

CHAPTER 6 – Application Examples

Tunable Filter and Oscillator using Mixers/Multipliers

In a) below, the filter can be configured as a lowpass, band pass, notch, or highpass and is tunable over a 10:1 range based on the tuning voltage

In b) below, the oscillator is tunable over a 10:1 range based on the tuning voltage.

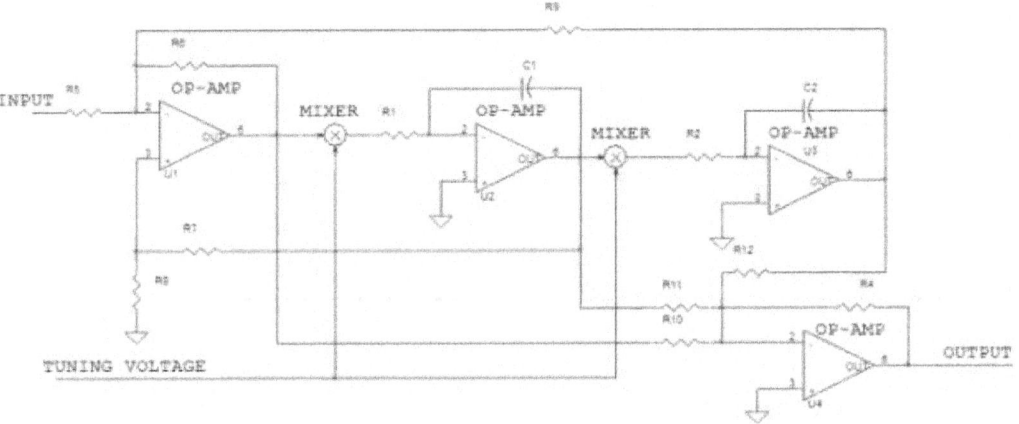

Figure 44: Tunable Filter using Multipliers/Mixers

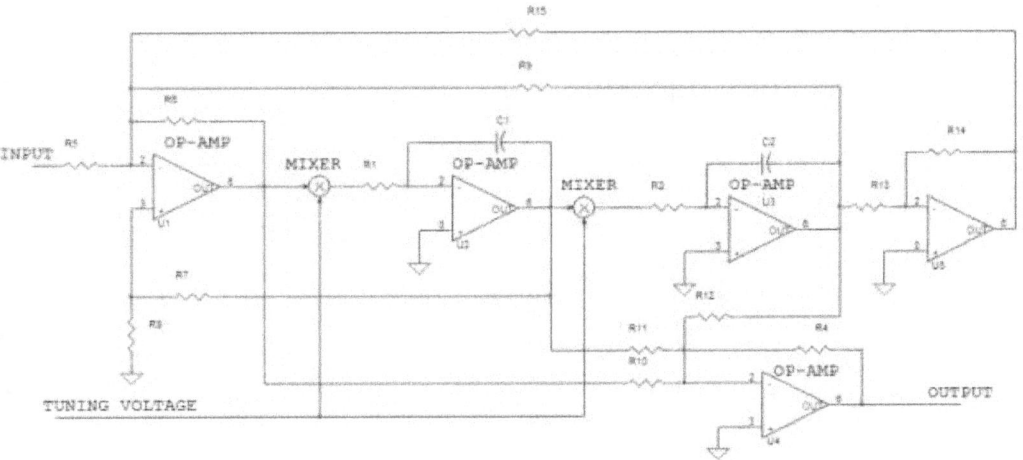

Figure 45: Tunable Oscillator using Multipliers/Mixers

Multiplying Circuits

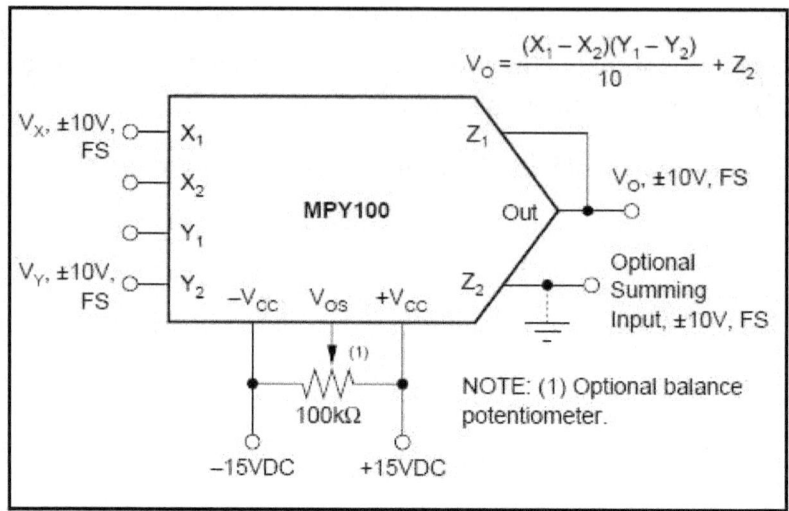

$$V_O = \frac{(X_1 - X_2)(Y_1 - Y_2)}{10} + Z_2$$

V_X, ±10V, FS — X_1

X_2

Y_1

V_Y, ±10V, FS — Y_2

MPY100

Out — V_O, ±10V, FS

Z_1

Z_2 — Optional Summing Input, ±10V, FS

$-V_{CC}$ V_{OS} $+V_{CC}$

(1)

100kΩ

NOTE: (1) Optional balance potentiometer.

−15VDC +15VDC

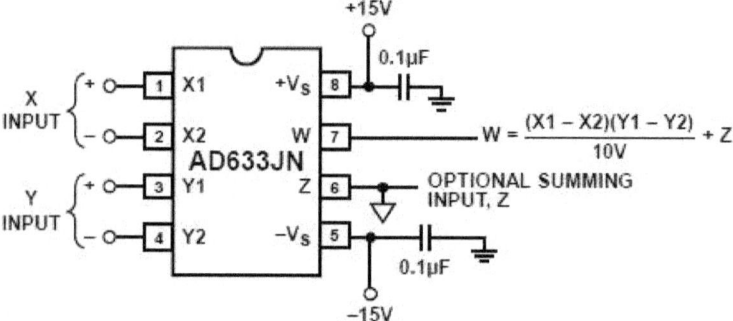

+15V

0.1µF

X INPUT { + — 1 X1   +V_S 8

− — 2 X2   W 7 } $W = \frac{(X1 - X2)(Y1 - Y2)}{10V} + Z$

AD633JN

Y INPUT { + — 3 Y1   Z 6 } OPTIONAL SUMMING INPUT, Z

− — 4 Y2   −V_S 5

0.1µF

−15V

Figure 46: Multiplication using Mixers

SQUARING

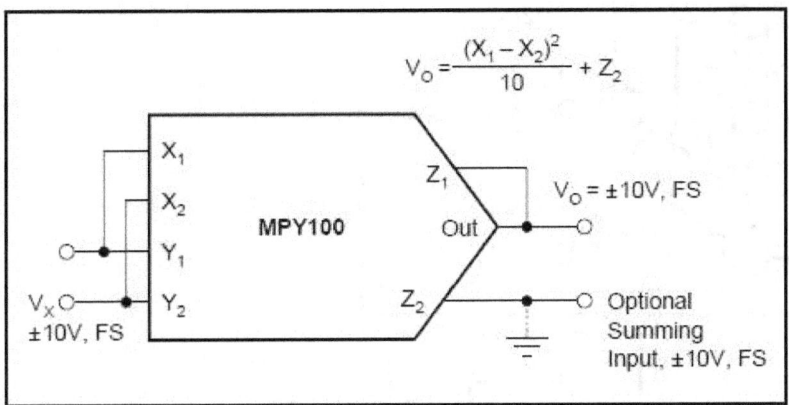

$$V_O = \frac{(X_1 - X_2)^2}{10} + Z_2$$

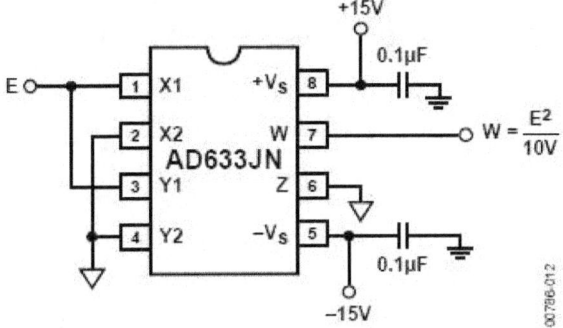

Figure 47: Squaring Circuits using Mixers

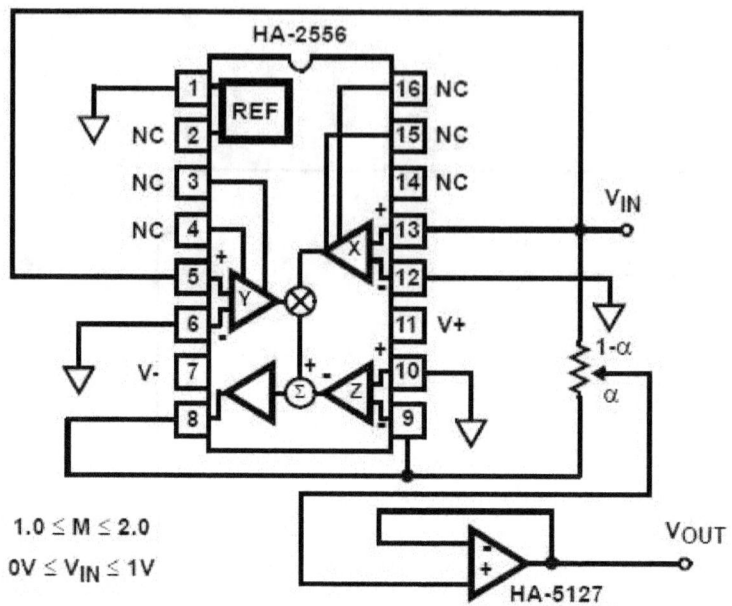

NONINTEGRAL POWERS - ADJUSTABLE

Figure 48: Mixer used to generate Non-integral Powers

AM Modulation Circuits

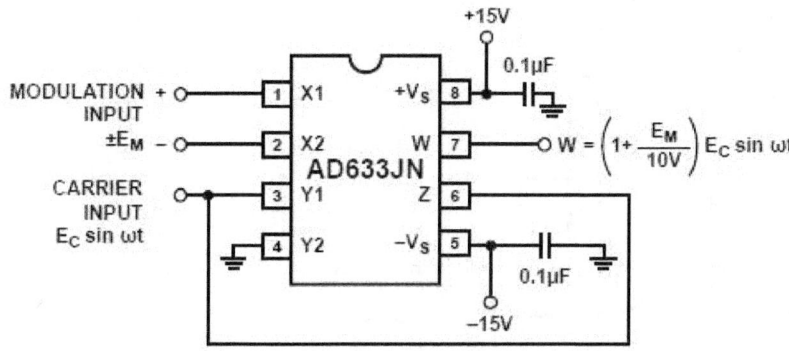

Figure 49: AM Modulation Circuit

Square Root Circuits

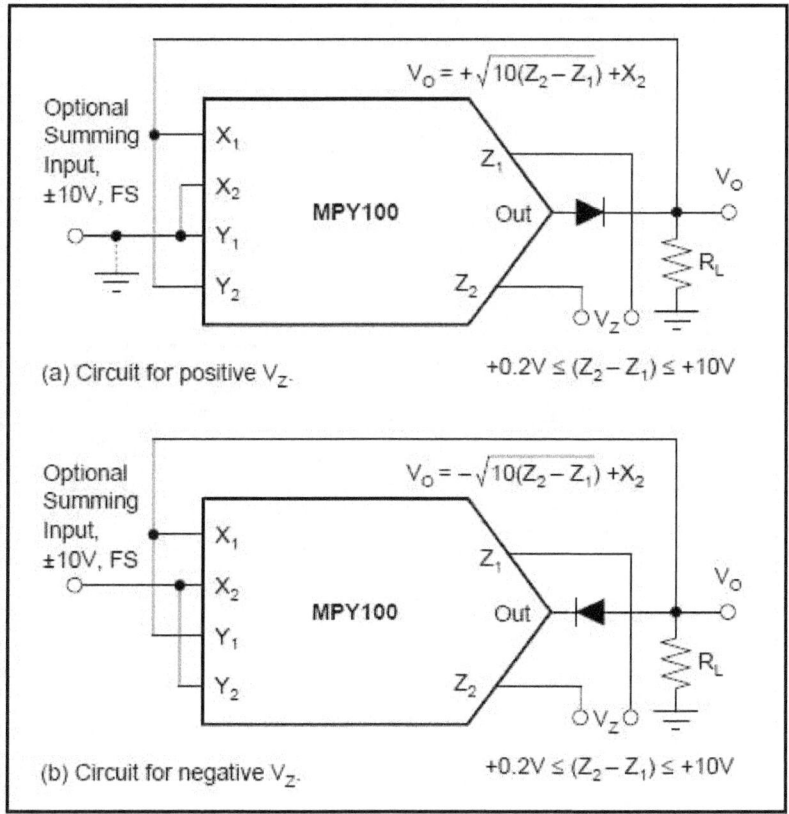

Figure 50: Square Root Circuit using MPY100 Mixer

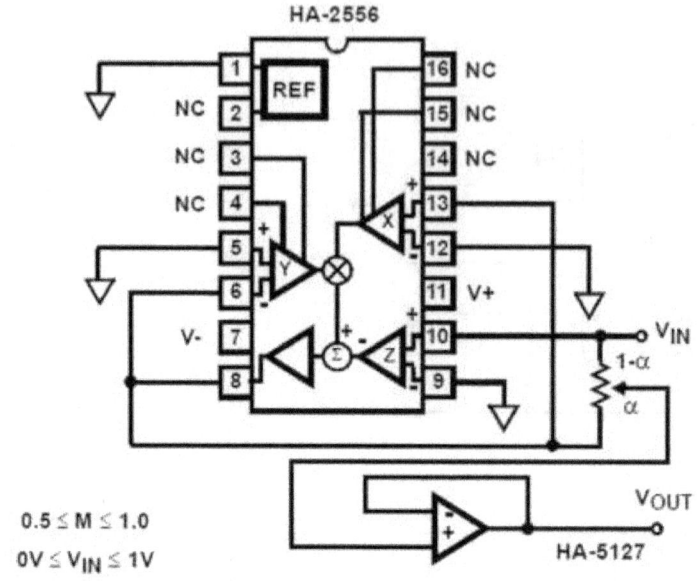

NONINTEGRAL ROOTS - ADJUSTABLE

$0.5 \leq M \leq 1.0$

$0V \leq V_{IN} \leq 1V$

Figure 51: HA-2556 Mixer used to product Non-Integral Roots Down Converters

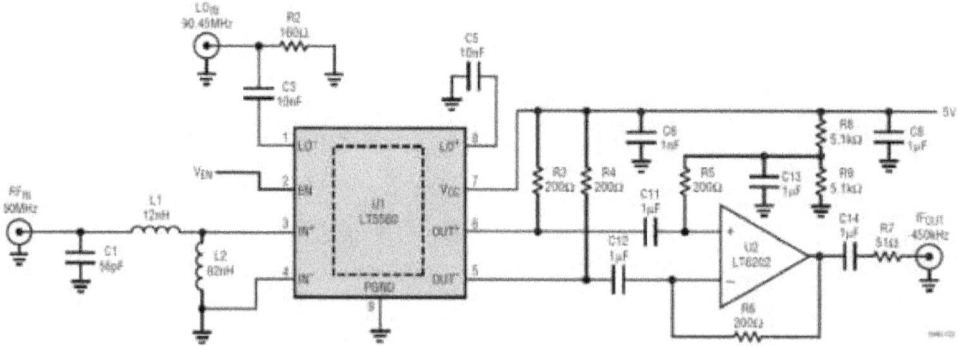

Figure 52: Down Converter Using LT5560 Mixer

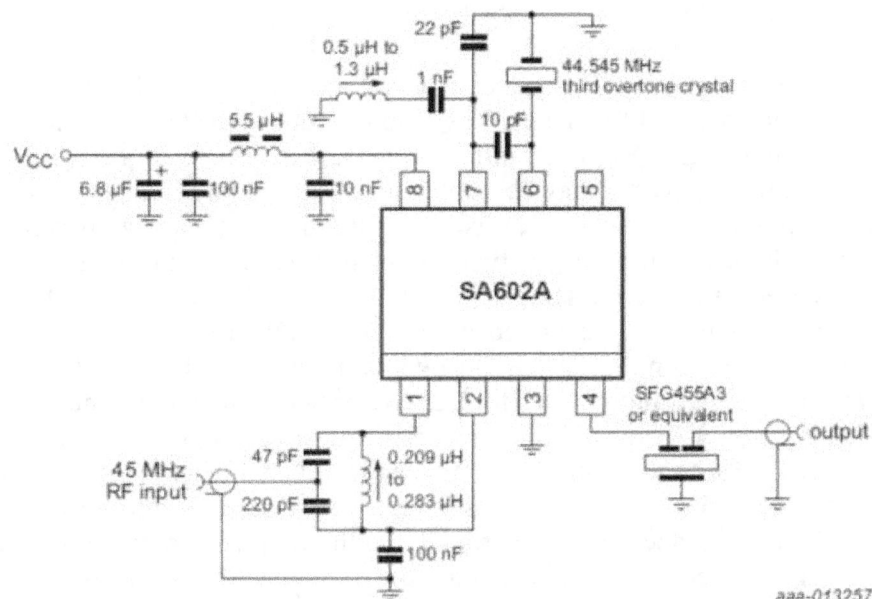

Figure 53: Down Converter using SA602A

aaa-013257

Down Converters using a Analog-to-Digital Converter

It is not well understood that down conversion can be performed using an ADC, by under sampling a signal frequency. The output frequency is the difference between the sample frequency (fS) (or a multiple of fS) and the incoming frequency. An analog-to-digital converter may be used to undersample any frequency that falls inside its full linear bandwidth. A mixer will produce a sum and a difference frequency. However, in a discrete time sampled system, the sum frequency and the difference frequency are the same apparent frequency. It is only the difference frequency that remains. In an undersampled system, the restriction is that the bandwidth of the incoming signals must be within the Nyquist zone. (A Nyquist zone extends over a bandwidth of fS/2, above or below an integral multiple of the sample frequency.) A signal falling outside the desired Nyquist zone wraps back into the DC-to-fS/2 zone. This constraint can be relaxed if subsequent band pass filtering in the digital domain restricts the frequency range of interest. As long as an undesired signal does not wrap around into the frequency range of interest, the effect on the desired spectrum is negligible.

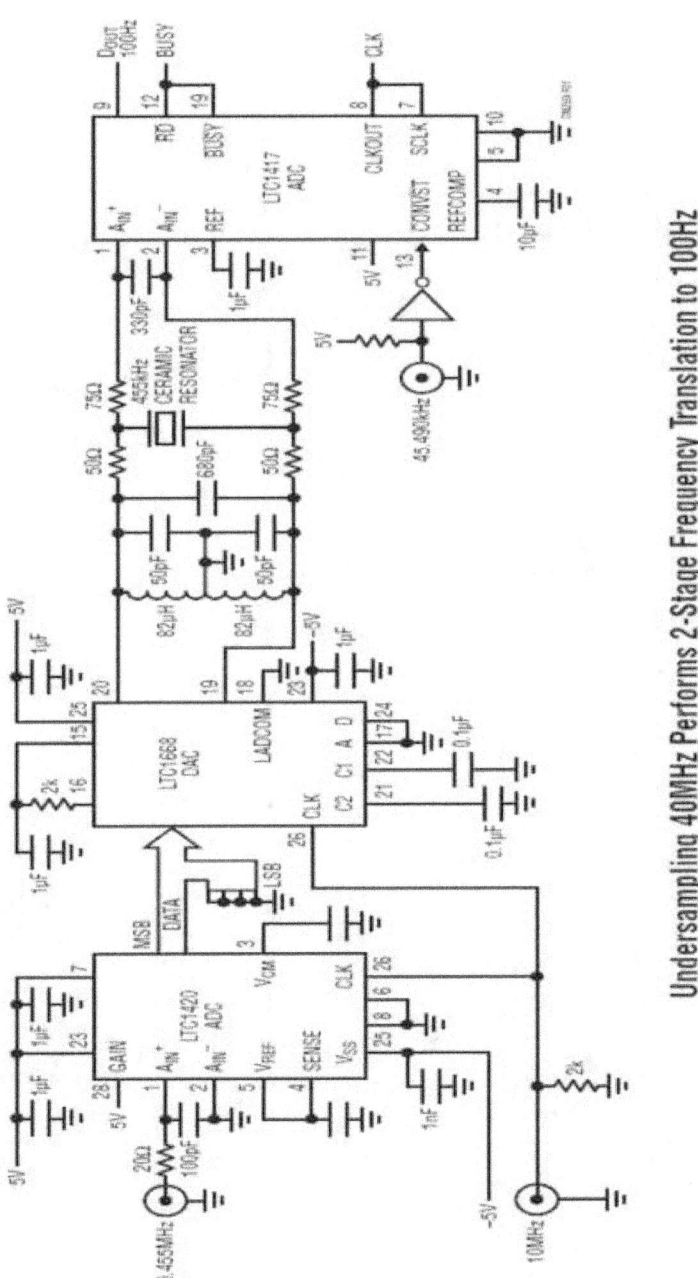

Figure 54: Analog-to-Digital Converter used as a Mixer for Down Conversion

CHAPTER 7 – Complete Circuits using Mixers

Transistor AM Radio using NPN 9018 as the Mixer

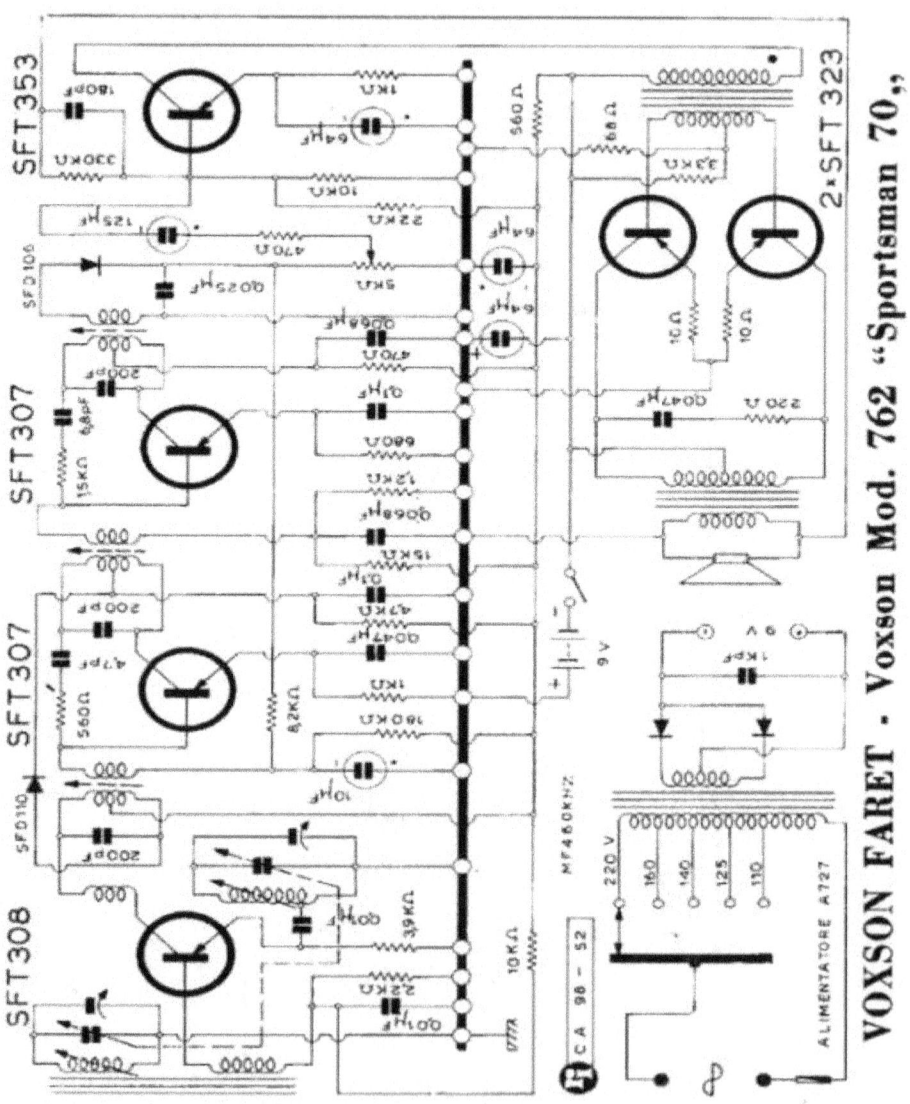

Figure 55: 9108 NPN Transistor as Mixer

Shortwave Receiver using SA602 Mixer

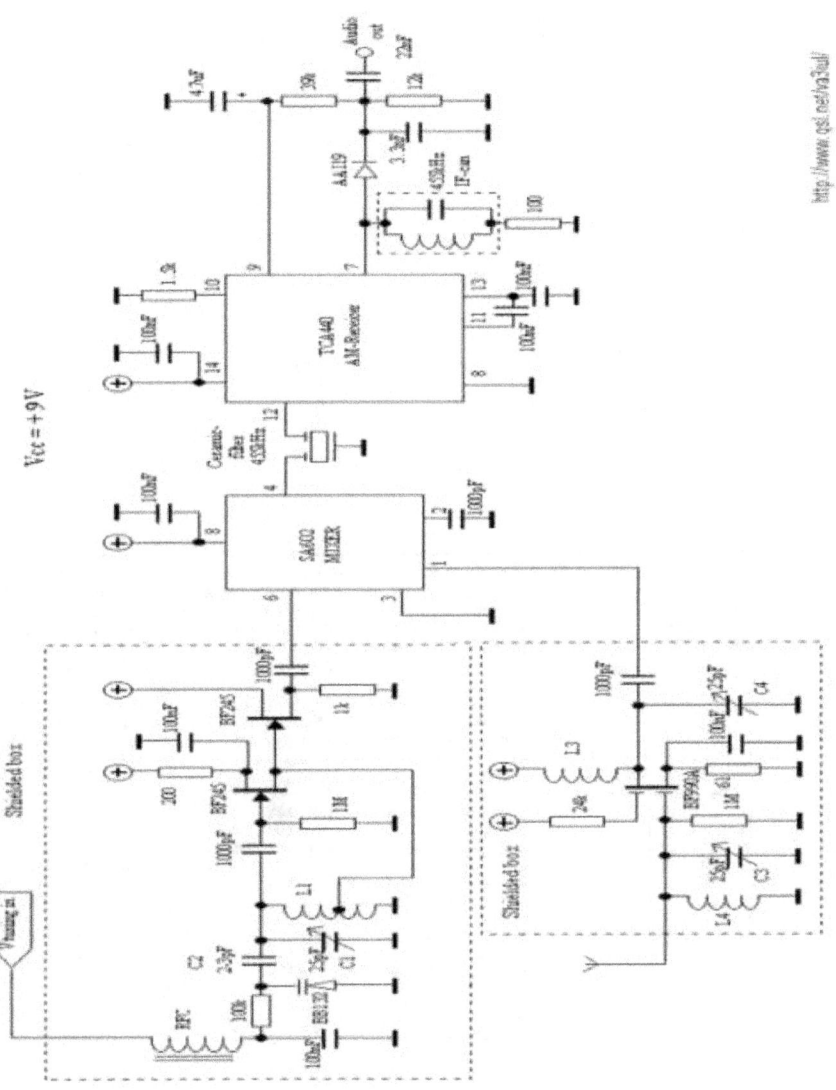

Figure 56: SA602 as Mixer Shortwave Receiver using SBL-1 as Mixer

Shortwave Receiver using SBL-1 as Mixer

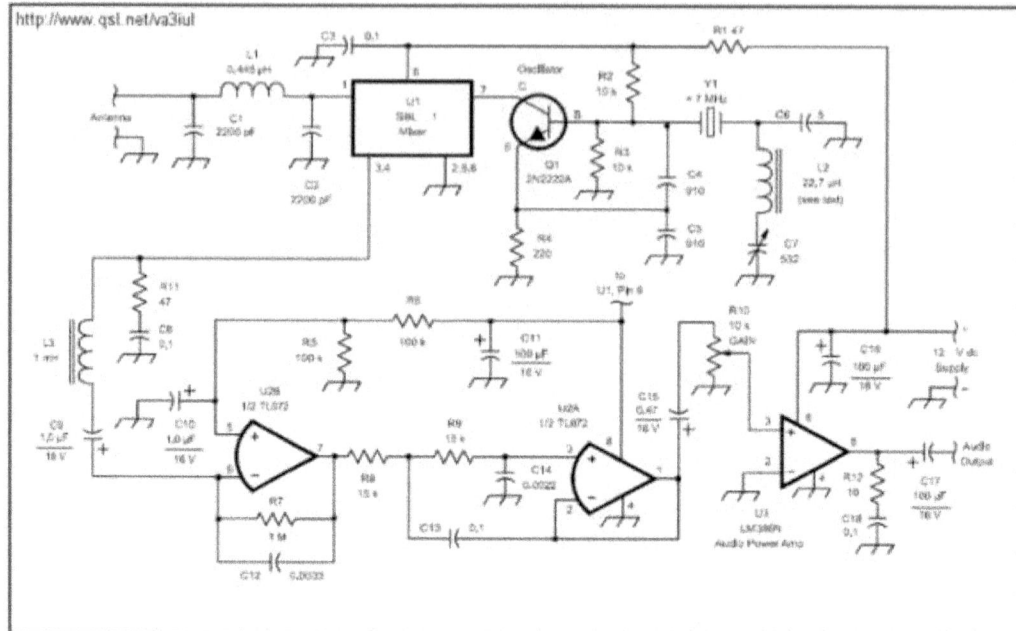

Fig 12.9 — An SBL-1 mixer (U1, which contains two small RF transformers and a Schottky-diode quad), a TL072 dual op-amp IC (U2) and an LM386 low-voltage audio power amplifier IC (U3) do much of the Rock-Bending Receiver's magic. Q1, a variable crystal oscillator (VXO), generates a low-power radio signal that shifts incoming signals down to the audio range for amplification in U2 and U3. All of the circuit's resistors are ¼ W, 5% tolerance types; the circuit's polarized capacitors are 16 V electrolytics, except C10, which can be rated as low as 10 V. The 0.1 μF capacitors are monolithic or disc ceramics rated at 16 V or higher.

C1, C2 — Ceramic or mica, 10% tolerance.
C4, C5, and C6 — Polystyrene, dipped silver mica, or C0G (formerly NP0) ceramic, 10% tolerance.
C7 — Dual gang broadcast variable capacitor (14-380 pF per section). ¼ inch dia shaft, available as #BC13380 from Ocean State Electronics. A rubber equipment foot serves as a knob. (Any variable capacitor with a maximum capacitance of 350 to 600 pF can be substituted; the wider the capacitance range, the better.)
C12, C13, C14 — 10% tolerance. For SSB, change C12, C13 and C14 to 0.001 μF.
U2 — TL072CN or TL082CN dual JFET op amp.

L1 — Four turns of AWG #18 wire on ¾ inch PVC pipe form. Actual pipe OD is 0.85 inch. The coil's length is about 0.65 inch; adjust turns spacing for maximum signal strength. Tack the turns in place with cyanoacrylic adhesive, coil dope or RTV sealant. (As a substitute, wind 8 turns of #18 wire around 75% of the circumference of a T-50-2 powdered-iron core. Once you've soldered the coil in place and have the receiver working, expand and compress the coil's turns to peak incoming signals, and then cement the winding in place.)
L2 — Approximately 22.7 μH; consists of one or more encapsulated RF chokes in series (two 10-μH chokes [Mouser #43HH105 suitable] and one 2.7-μH

choke [Mouser #43HH276 suitable] used by author). See text
L3 — 1 mH RF choke. As a substitute, wind 34 turns of #30 enameled wire around an FT-37-72 ferrite core.
Q1 — 2N2222, PN2222 or similar small-signal, silicon NPN transistor.
R10 — 5 or 10 kΩ audio-taper control (RadioShack No. 271-215 or 271-1721 suitable).
U1 — Mini-Circuits SBL-1 mixer.
Y1 — 7 MHz fundamental-mode quartz crystal. Ocean State Electronics carries 7030, 7035, 7040, 7045, 7110 and 7125 kHz units.
PC boards for this project are available from FAR Circuits.

Figure 57: Minicircuits SBL-1 as Mixer

Receiver using Dual Gate MOSFET as Mixer

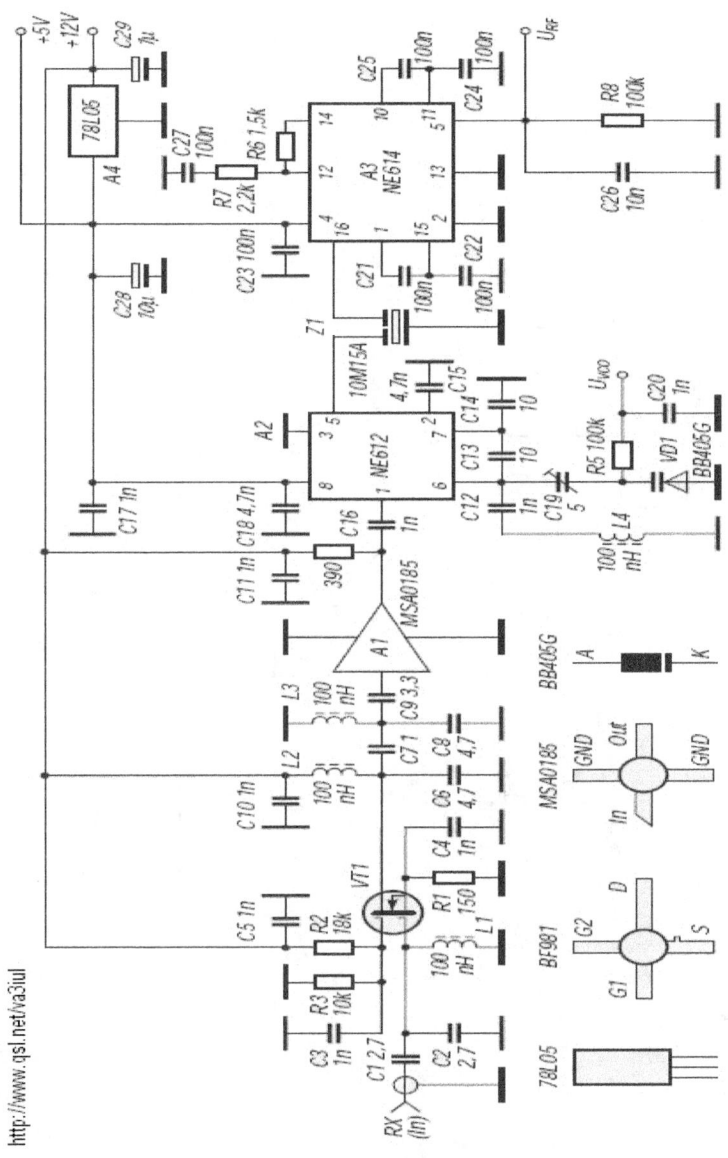

http://www.qsl.net/va3iul

Figure 58: Dual Gate MOSFET as Mixer in 144MHz FM Radio

Receiver using ADE-1 as Mixer

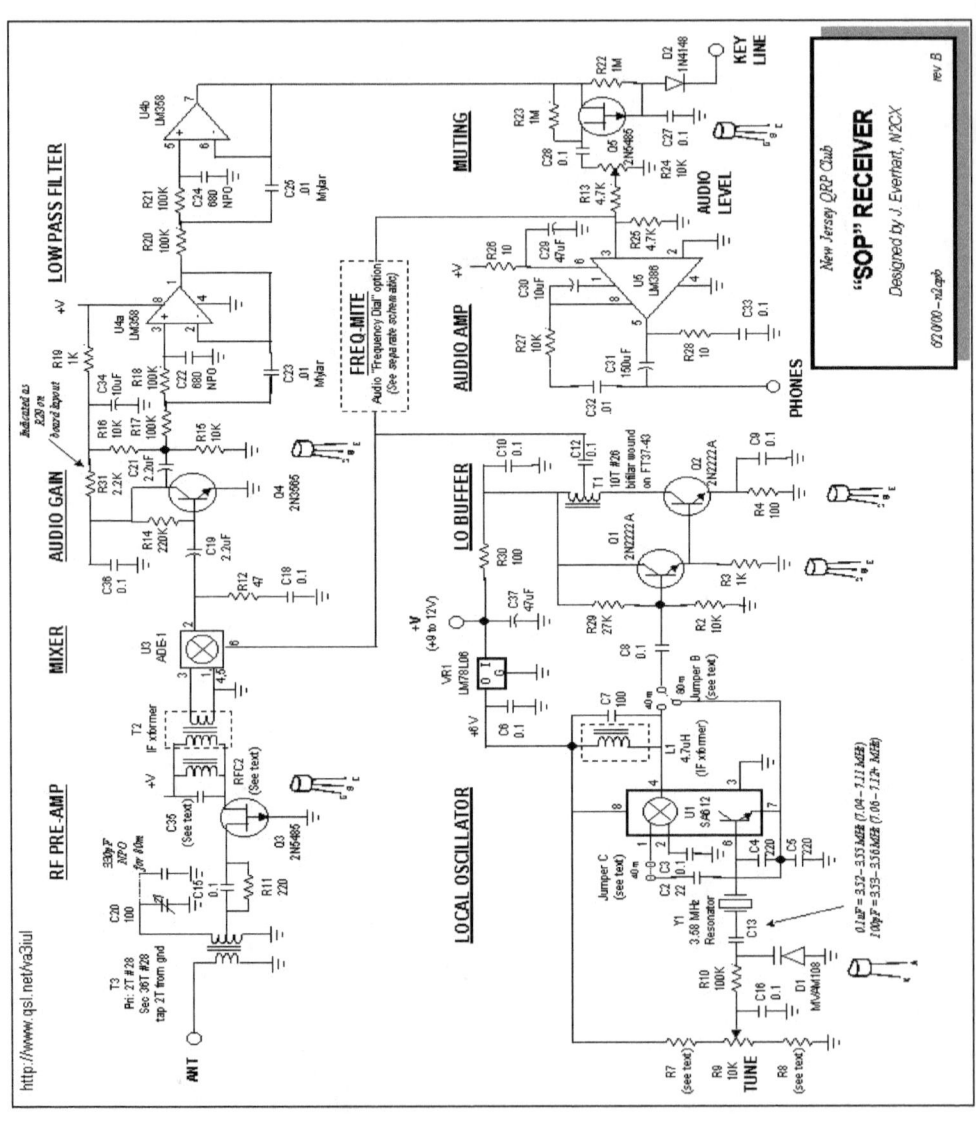

Figure 59: Minicircuits ADE-1 as Mixer

Spectrum Analyzer Block Diagram and Schematic

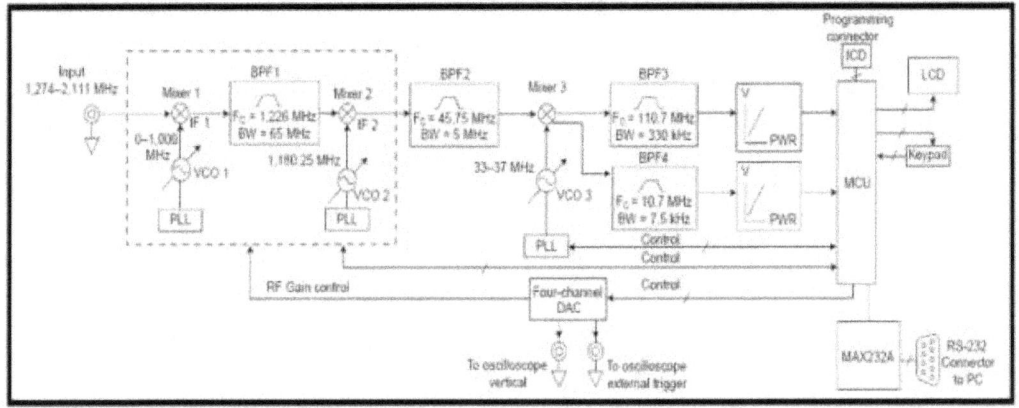

The MAX3550 performs many of the spectrum analyzer's front-end functions. The SA612A mixer shifts the received signal to 10.7 MHz for filtering. The AD8307 amplifier performs the conversion to decibels relative to 1 milliwatt (dBm).

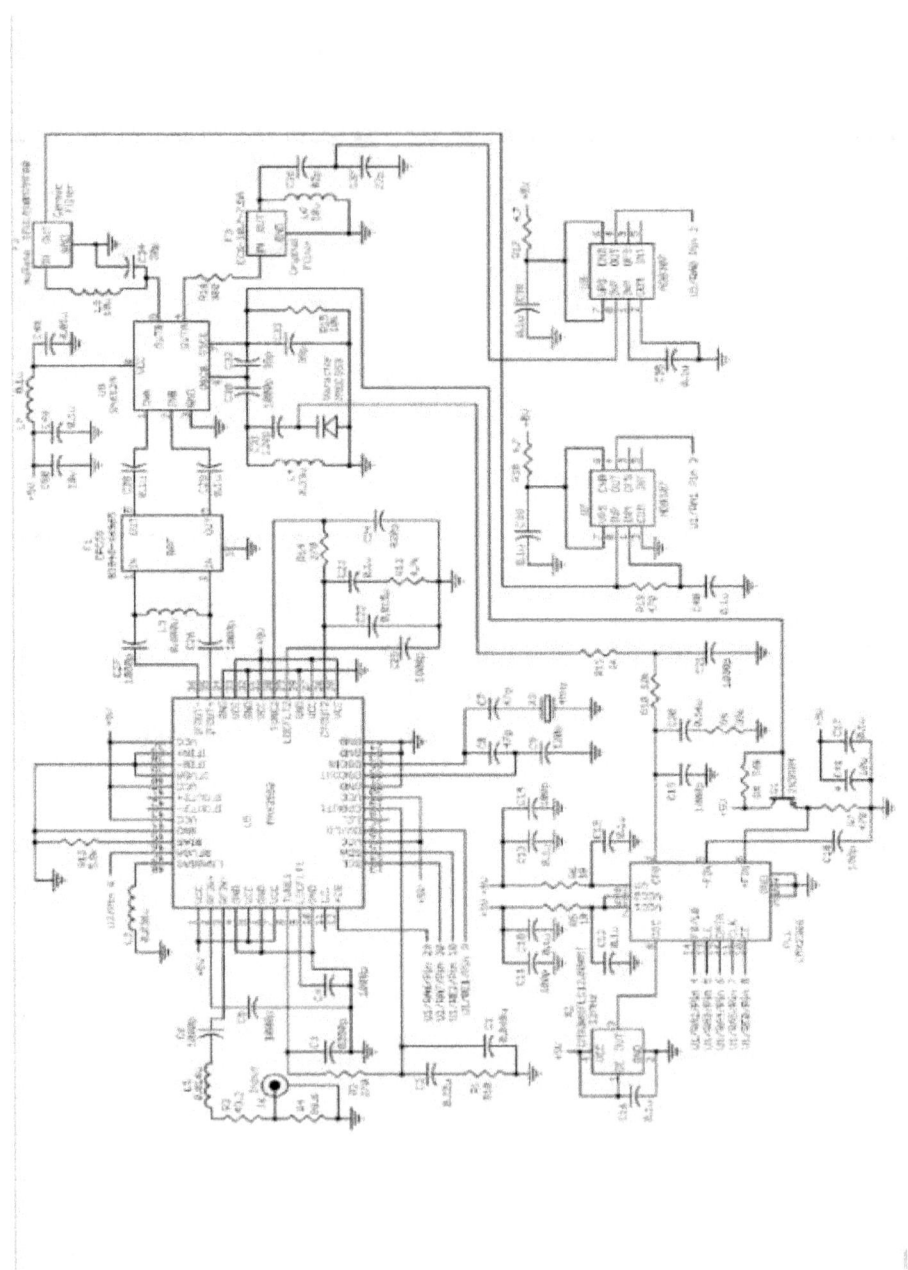

Figure 60: Spectrum Analyzer Block Diagram and Schematic.

The *ARRL Handbook for Radio Amateur*, (American Radio Relay League, Newington, 1993), and many previous editions.

Joseph J. Carr, *Mastering Radio Frequency Circuits Through Projects & Experiments*, McGraw-Hill, 1994

The Watkins-Johnson Company, *RF and Microwave Designer's Handbook*, 1993

Saraga, W., "The Design of Wide-Band Phase Splitting Networks'," Proc IRE, Vol 38, p 754 (1950)

Cayley, A., *An Elementary Treatise on Elliptic Functions,* (Bover, New York, 1961).

Abramowitz, M., and Stegun, I., *Handbook of Mathematical Functions with Formulas, Graphs, and Mathematical Tables*, National Bureau of Standards, Applied Mathematics Series, (US Government Printing Office, Washington, DC 1964

Schmidt, Kevin, Phase-Shift Network Analysis and Optimization

Cracknell, A.P., Applied Group Theory, (Pergaman, Oxford, 1968)

www.ingramcontent.com/pod-product-compliance
Lightning Source LLC
Chambersburg PA
CBHW070935180526
45168CB00003B/1086